Editors

Doru Costache (ISCAST, Sydney College of Divinity)
Mark Worthing (ISCAST)

Book Reviews Editor

David James Hooker (ISCAST, Monash University)

Advisory Board

Denis Alexander (ISCAST, Faraday Institute)
Andrew Briggs (University of Oxford)
Ted Davis (Messiah University)
Peter Harrison (ISCAST, University of Queensland)
Nicola Hoggard Creegan (New Zealand Christians in Science)
Roger Lewis (ISCAST, University of Wollongong)
Victoria Lorrimar (ISCAST, Trinity College Queensland)
Graeme McLean (ISCAST, Charles Sturt University)
Tom McLeish (ISCAST, York University)
Neil Ormerod (ISCAST, Alphacrucis College)
John Pilbrow (ISCAST, Monash University)
Gillian Straine (Faraday Institute)
Lisanne Winslow (University of Northwestern)
Jennifer Wiseman (ISCAST, NASA)

Editorial Committee

Katherine Bensted (St Peter's College Adelaide)
Jackie Liu (ISCAST)
Chris Mulherin (ISCAST, University of Divinity)

For submissions: editor@iscast.org
For general enquiries: contact@iscast.org

Christian Perspectives on Science and Technology

The ISCAST Journal

ISSN 2653-648X (Online)
ISSN 2653-7656 (Print)

https://journal.iscast.org/
New Series, 1, 2022

About the Journal

The ISCAST journal, *Christian Perspectives on Science and Technology* (*CPOSAT*), was relaunched in 2022. Capitalising on the previous years of publication (online since 2006; its rich archive is available on the ISCAST website), the journal enters a new stage of life with its relaunch as a world-standard academic resource.

The ISCAST journal is unique in the Australian landscape and one of the few journals globally that discusses the nexus of science, technology, faith, ethics, and spirituality. In doing so, it advances ISCAST's mission of promoting a climate of mutual understanding and constructive exchange between science and technology practitioners, and people of faith.

The target readership includes academics interested in science and faith, as well as educators, church leaders, and postgraduate and graduate students.

The relaunched journal is an online, open-access resource, inviting original contributions from national and international scholars. It publishes book reviews and double-blind peer-reviewed articles. The accepted articles and book reviews are published as they become available. At the closing of each annual edition, the published materials are collected in one document, also made available via the journal's website.

We especially invite proposals for articles in science/technology that have theological/ethical/spiritual implications, and articles in theology/ethics/spirituality that engage scientific/technological topics. Original studies of the history of science and faith are equally welcome.

While the authors retain the copyright for their respective works, the materials published in *CPOSAT* may be freely disseminated, with due acknowledgment of their authorship and the place of original publication.

Information for authors
https://journal.iscast.org/submit-an-article
https://journal.iscast.org/submit-a-book-review

CPOSAT is indexed with CrossRef.
https://doi.org/10.58913/isxa

Contents

Editorial ix

Note from the ISCAST Executive Director xi

Articles

Genesis 1–3 as a Resource for Twenty-First Century Faith
Carolyn M. King 1

Between Darwin and Dostoevsky:
The Syntheses of Theodosius Dobzhansky
Christopher Howell 28

From Physics to Metaphysics: A New Way
Stephen Ames 46

Faith, Deuteronomy 18:21–22, and the Scientific Method
Charles Riding 72

Bruce Craven: Contrarian or Questioning Thinker?
John Pilbrow 96

On Subjects, Objects, Transitional Fields, and Icons:
The Semiotics of a New Paradigm in Human Studies
Marcello La Matina 108

Evolution as History: Phylogenetics of Genomes and Manuscripts
Graeme Finlay 150

Imago Dei in the Age of Artificial Intelligence:
Challenges and Opportunities for a Science-Engaged Theology
Marius Dorobantu 175

Book Reviews

Jacob Shatzer: *Transhumanism and the Image of God*
Andrew Sloane 197

Ronald L. Numbers (ed.): *Galileo Goes to Jail: And
Other Myths about Science and Religion*
Robert Brennan 201

James Ungureanu: *Science, Religion, and the Protestant
Tradition: Retracing the Origins of Conflict*
Robert Brennan 202

Ronald E. Osborn: *Death Before the Fall: Biblical
Literalism and the Problem of Animal Suffering*
Andrew Sloane 205

Graeme R. McLean: *Ethical Basics for the Caring Professions:
Knowledge and Skills for Thoughtful Practice*
Denise Cooper-Clarke 208

Graeme R. McLean: *Ethical Basics for the Caring Professions:
Knowledge and Skills for Thoughtful Practice*
Tom Slater 212

Mike Hulme: *Why We Disagree About Climate Change:
Understanding Controversy, Inaction and Opportunity*
Richard Gijsbers 215

David Frost: *Blind Evolution? The Nature of
Humanity and the Origin of Life*
Peter Stork 219

Ian Hutchinson: *Can A Scientist Believe in Miracles? An MIT Professor Answers Questions on God and Science*
D. Gareth Jones 223

Simon Conway Morris: *From Extraterrestrials to Animal Minds: Six Myths of Evolution*
Jonathan Clarke 226

Graham Buxton and Norman Habel (eds): *The Nature of Things: Rediscovering the Spiritual in God's Creation*
Jonathan Clarke 230

Robert John Russell & Joshua M. Moritz (eds): *God's Providence and Randomness in Nature: Scientific and Theological Perspectives*
Neil Ormerod 235

Suzie Sheehy: *The Matter of Everything: Twelve Experiments that Changed our World*
Charles Sherlock 237

Graeme Finlay: *Evolution and Eschatology: Genetic Science and the Goodness of God*
D. Gareth Jones 240

Editorial

In this age of intense dissemination of information, a new academic journal is not what one would call a rare bird. So why another new journal? Why, then, the relaunch of *Christian Perspectives on Science and Technology* (*CPOSAT*)?

ISCAST, the publisher of this journal, has as its mission "to engage Australians in constructive conversation between Christian faith and the sciences." In fostering this conversation—which has international ramifications, as evidenced by its cooperation with New Zealand Christians in Science and other organisations and people beyond Australia—ISCAST is guided by its core values: commitment to Christian foundations, scientific integrity, and the desire to offer a theological and an academic safe space.

As a part of fulfilling its mission in the light of its core values, ISCAST is engaged in a number of undertakings, including the relaunch of *CPOSAT*. This is the answer to our earlier question.

In closing this inaugural edition of the journal's new series, the editors look back at the tremendous work of this year with satisfaction and joy. We are thankful to God for blessing this ministry of ISCAST and we are grateful to all who have supported it—from the distinguished members of the Advisory Board and the Book Reviews Editor, David Hooker, to the tireless Editorial Committee, the authors, the reviewers, the proofreaders, and the typesetters. Last but not least, we are grateful to the Australian Research Theology Foundation, Inc, whose generous grant made possible the creation of the journal's online platform. And, while work on the website is still underway, we all benefit from the expediency of having at our disposal a modern, user-friendly means of disseminating and accessing information. We also take this opportunity to express our gratitude to Jackie Liu, who maintains the website and secures the wonderful typesetting of the published materials.

Many readers would be familiar by now with the content of this inaugural edition of *CPOSAT*, as the published materials have been readily available to the public on the website. That said, it is our pleasure to offer the full edition in the form of one document.

Before we sign off, inviting the reader to browse the journal, we would like to highlight the international scope of *CPOSAT*, obvious in the contributions published here. The eight articles are written by authors from Aotearoa New Zealand (two), Australia (three), Italy (one), the Netherlands (one), and the United States of America (one). Also noteworthy is that the authors represent several Christian traditions: Anglican, Catholic, Evangelical, and Orthodox. ISCAST is pleased to provide a space that fosters creative encounters for the purposes of edifying readers, for God's glory.

Doru Costache
Mark Worthing

December 2022

Note from the ISCAST Executive Director

It is a great pleasure to add my word of thanks to the editors (as well as all those mentioned previously) for this sterling production. What an impressive feat for a small not-for-profit Christian organisation in the Antipodes! Thank you Doru, Mark, and Dave and all your helpers for this great effort in further promoting a constructive conversation between the sciences, technology, and robust Christian faith.

It is also my pleasure to announce that three "ARTFInc awards" of $500 each are in order. Congratulations to Marius Dorobantu, Carolyn King, and Graeme Finlay. The judging panel found it hard to choose three of the articles in this volume; however, these three authors have been chosen for the prizes. Our thanks go again to the Australian Research Theology Foundation Inc., which not only supported the new website construction but also provided these prizes.

We look forward in coming years to continue playing our part in glorifying Jesus Christ through this journal and the wider work of ISCAST.

Chris Mulherin
ISCAST Executive Director

Genesis 1–3 as a Resource for Twenty-First Century Faith

Carolyn M. King

Abstract: Centuries of pre-scientific tradition underlie the widespread modern misunderstanding of the Book of Genesis. But, in fact, it is full of sharply relevant wisdom for the here and now. We can find real inspiration when we read it with attention to the original cosmological environment of Genesis 1, which supports the idea that it is not a prehistoric account of the origins of the universe, but the text of a six-day festival celebrating the inauguration of the cosmos as a fit and functional home for humanity. Likewise, a contemporary reinterpretation of the Eden story of Genesis 2–3 in terms of the origins, anatomy, and functions of the human brain can undo millennia of guilt and grief imposed by the idea of original sin. In this light, a serious, respectful, and integrated approach to Genesis based on the best of biblical scholarship and of modern neurobiology can reassure us that the widely assumed warfare between science and religion was never necessary in the first place. Rather, a deeply informed biblical faith can inspire us with new confidence in God and in our own human nature.

Keywords: ancient cosmology; six-day creation; Eden myth; science and religion; contemporary faith

It could be argued that centuries of misunderstanding of the Old Testament is the single most significant cause of the supposed warfare between science and religion. The long-standing warfare image is aggravated by the way that the most strident voices from either party rarely

---

Carolyn M. King, FRSNZ, is Professor Emerita, University of Waikato, Hamilton, New Zealand. She holds doctorates in science (Oxford, 1971) and religious studies (Waikato, 1999). For a complete bibliography and list of research awards, go to https://www.waikato.ac.nz/staff-profiles/people/cmking.

admit their mutual lack of training in the sophisticated philosophy and serious literature underlying their opponents' position, so neither can see how far each so dangerously underestimates the other. The primary message of this paper is that ancient and contemporary knowledge are better read as cooperating in advancing our understanding of ourselves. This is news that we must break to our contemporaries, especially to students.

Nothing inflames the conflict faster than derogatory criticism of real science by religious believers steeped only in naïve misreadings of the Book of Genesis,[1] opposed by arrogant rejection of all forms of faith from scientists with no knowledge of biblical scholarship.[2] Both sides depend on arguments based on simple, oft-repeated errors of fact, logic, and interpretation, and of basic scholarship. In turn, careful attention to the real bases of both disciplines shows that nearly all apparent contradictions are illusionary.

The Book of Genesis was not written as a single narrative. During its most formative centuries, its content had never been written down at all. It is the product of long, slow years of development of ancient oral traditions dating back to at least 1200–1000 BCE, through multiple generations of people who could not read or write but had phenomenal memories. So the text as we have it is a composite of independent oral and written traditions and complementary points of view.[3] The two creation stories preserved in the Book of Genesis have very different histories and backgrounds. That means that they must be read and understood differently.

---

[1]   R. L. Numbers, *The Creationists: The Evolution of Scientific Creationism* (Berkeley: University of California Press, 1992) provides a comprehensive history of creationism. J. C. Sanford, *Genetic Entropy & the Mystery of the Genome* (New York: FMS Publications, 2008) updates creationist views on advances in genetics.

[2]   R. Dawkins, *The God Delusion* (London: Bantam Press, 2006). C. Hitchens, *God Is Not Great: The Case against Religion* (USA: Hachette Book Group, 2007).

[3]   K. Armstrong, *A History of God: The 4,000 Year Quest of Judaism, Christianity, and Islam* (London: William Heinemann, 1993).

Genesis 2-3

The oldest written version of the oral tradition, dating to 950–850 BCE, is preserved in Genesis 2–11. This is the Primeval History of all humankind, concerned with all peoples because it long antedated the development of nations, and it describes the creation of the first humans, the Flood, and the new beginning after the Flood. To understand it, we must step into the worldview of the people of that time, not impose ours on them.[4]

The story of the Garden of Eden is a myth in the proper sense, that is, a story about human origins that is not itself literally true, but has significant truth in it. By updating the metaphorical images it uses, and expressing them in terms compatible with contemporary research, the traditional story can still be understood to have important things to say about modern human nature. One of the most profound of such insights concerns ancient ideas about how we make decisions, and especially about the origins of human social behaviour.

The Social Nature of Humanity

Anthropology has amply confirmed that human social behaviour has evolved in gradual stages from our primate ancestors.[5] For thinking people, whether religious or not, this is no longer a contentious issue, but the technical definition of humanity remains difficult, since human characteristics appeared slowly and over a succession of descendent species. The scientific story started with the origins of the earliest sociable anthropoids (the monkey/ape lineage) about 35 million years ago. The separation of the human lineage from the apes was complete by about 5 million years ago; the development of agriculture, the end of purely genetically based evolution, and the rise in influence of cultural traditions began between about 30,000 and 10,000 years ago.

---

4   J. H. Walton, *The Lost World of Genesis One: Ancient Cosmology and the Origins Debate* (Downers Grove, IL: InterVarsity Press, 2009).
5   J. K. McKee, F. E. Poirier, and W. S. McGraw, *Understanding Human Evolution*, 5th edn (Routledge, 2004).

Sociality always has been and still is as much a part of the definition of being human as is bipedal gait and a large brain, and it preceded both those characteristics by many millions of years. Not all primates are sociable, of course, but it is virtually certain that all species of humans and of their immediate ancestors, the australopithecines, always have been. Therefore, the philosopher Hobbes' assumption, that people are independent human egoists who make solitary decisions about social life, was simply wrong; the world never was full of independent human egoists. For the whole of the 4-5 million years or so that hominids have been evolving, and for some 30 million years before that during which the anthropoid ancestors of the human line were evolving, there has been no such thing as a solitary independent individual, except maybe a dead one. The idea that "the sheer dangers of anarchy had forced beings who were natural solitaries to make a reluctant bargain"[6] is based on a series of spectacular misunderstandings of the lives, minds, and social relationships of our human ancestors and of the sociable primates that preceded them.

Any characteristic which, like sociality, has been ingrained in our nature fully as deeply and for much longer than our large brains must exert a powerful influence over our lives. Morality is a key part of the problem of understanding human relationships, and sociality is necessarily linked to morality and to its opposite, the idea of sin. If we wish to understand the processes that have for millennia shaped the human mind and spirit, expressed in the conflicts arising within and between our social groups, we must first understand the processes that shaped the human species. As Mary Midgley puts it:

> Once we accept our evolutionary history as a general background, it is quite natural and proper to use it in explaining many elements of human life. If we shut morality off from that explanatory pattern of thought, we tend to make its relation to the rest of human life unintelligible, which cannot be an advantage.[7]

---

6   M. Midgley, *The Ethical Primate: Humans, Freedom and Morality* (London: Routledge, 1994), 110.

7   Midgley, *The Ethical Primate*, 14.

Of course, that is not to say that what is natural is necessarily good. There is no need to adopt the ruthless values of natural selection as our own. But if we decide to develop values that are different from those favoured in our ancestors by natural selection, or we wish to change some disconcertingly stubborn parts of our nature, we need to know what we're up against.[8] The advance of medical science has offered solutions to many old questions about how our minds work, although the details are perpetually controversial.[9]

Mind and Brain

First, there is a difference between mind and brain. Although our mind is located in parts of our brain, the brain itself is only a physical organ, whereas the mind is a coordinated set of thinking faculties and reasoning processes including consciousness, imagination, perception, thinking, judgement, language, memory, and emotions. Brain and mind are connected through neural pathways transmitting signals controlling our everyday functions, from breathing, digestion, and pain sensations to movement, thinking, and feeling, and the making of moral judgements.

Evolutionary psychology recognises morality as a product of natural selection, just as is any physical feature. Wright points out that the similarity in physique that makes every page of *Gray's Anatomy* applicable to all humans of all races applies also to their mental architecture—the basic structure of the human mind is species-typical. It is therefore reasonable to speak of "the psychic unity of humankind."[10]

Second, we can now understand the complex structure and long evolution of our physical brains. Far from being a simple box into which teachers can dump information, the human brain is a complicated structure of three main parts, each of which has a different history and set of functions. Only when we appreciate how differently these

---

8   R. Wright, *The Moral Animal* (New York: Random House, 1994), 31.
9   Rather than trying to cite any particular source for this statement, a simple internet search will reveal some of the many ideas under current discussion.
10   Wright, *The Moral Animal*, 26.

three parts develop, operate, and interact with each other, can we begin to understand how our mind works. More importantly, we can see how that understanding can underpin many of our most ancient perceptions of ourselves, and our internal dilemmas and moral conflicts.

The Snake in Our Heads

The brain comprises three distinct parts, with different evolutionary origins and contemporary functions. Understanding how this complex structure evolved, how its apparently independent parts work together so perfectly, and the implications of this seamless integration for religious belief, can suggest a new set of contemporary metaphors that revolutionise traditional interpretations of Genesis 2–3.[11] Like all metaphors, this one has its limitations, but it is certainly a clear example of the huge significance of intimate communication between three parts comprising one holistic body. This is not a new idea for Christians. Augustine taught that humanity is created, not merely in the image of God, but in the image of the Trinity. So, by the grace of God, we can humbly model the Trinity in our own experience of the three components of our brain as different but loving and completely interdependent parts of our own minds.

The Hindbrain

This is the most ancient component of the brain, lying tucked underneath the main structure, where the top of the spinal cord reaches the base of the brain proper. It controls all our unconscious processes, like breathing, digestion, balance, sleep cycles, visual processing, heart, and circulation. All our most powerful and ancient urges, long needed to satisfy our ancestors' needs for food, sex, and flight from danger, start from here. The very same structures can be found throughout our lineage, dating back to the earliest vertebrates of 450 million years ago.

---

11 M. Dowd, *Thank God for Evolution: How the Marriage of Science and Religion Will Transform Your Life and Our World* (New York: Viking, 2008), 149.

The genes controlling these processes have been copied down the generations through all our ancestral forms, from before the armoured fishes of the Palaeozoic,[12] through the earliest tetrapods (four-legged animals) to all the reptiles from the Mesozoic to the present. Mutations in genes managing such basic and indispensable functions were instantly fatal, which is why we have inherited them largely unchanged. They comprise the reptilian ancestry of many lower (i.e., automatic) functions of the brains of all later vertebrates, right down to people. The genes that make the scales that clothe the legs of living birds are probably much the same as those that did the same for their reptilian forebears.

The idea of such long-term constancy seems far-fetched, but in fact Nature is very conservative, and seldom invents a new process if a slightly modified old one still works. In human engineering we say "don't reinvent the wheel." Proof of it as applied to our own brains can be demonstrated from Shubin's eloquent account of the way the origins of the first ten cranial nerves that emerge from underneath the brains of sharks and dogfish, and run to the nose, eyes, ears, jaws, etc. are still exactly the same in number, origin and function in humans, although their pathways are substantially rearranged to fit in our differently shaped skull. People who still get hung up on the idea of humans being related to apes have no idea of how far they have underestimated the length of the real and much more wonderful story of our emergence from the lower animals.

The hindbrain is the origin of all our unconscious preferences to "Look after Number One"—our preprogrammed tendency to self-preservation, which conflicts with much of what our conscious education commands us to do. The feeling is well known to anyone seriously attempting to obey our higher moral imperatives. St Paul's oft-cited complaint hit the nail right on the head: "When I want to do the right, only the wrong is in my reach ... there is in my bodily members a different law, fighting against the law that my reason approves" (Rom 7:21–

---

12   N. Shubin, *Your Inner Fish: A Journey into the 3.5 Billion-Year History of the Human Body* (New York: Pantheon, 2008).

24). Paul could hardly have written a better description of the inner conflicts generated by the activities of the hindbrain if he had been schooled in medical science.

The Midbrain

Standard anatomical texts illustrate the position of the midbrain, buried in the middle of the cranial mass, above the top of the spinal cord and the hindbrain and below the forebrain, which lies on top of both. It is the seat of the limbic system, which includes several important glands which produce the hormones that race around the body in the bloodstream. They coordinate information from the senses and the muscles, and control many vital bodily functions.

The limbic system is among the products of the later evolutionary heritage of humankind. Reptiles don't have a limbic system, but all mammals do. The limbic system is important because it amplifies the unconscious signals from the hindbrain. It produces a great range of conscious emotions during waking and dreams during sleep, by adding feelings to basic urges, especially the need to find sexual contacts and compete with others for social status.

Feelings of love, fear, racial hatred, sexual jealousy, and many more that profoundly influence our daily decisions, are common to all people. The problem is that some of these run counter to moral wisdom. That introduces severe personal conflicts, because overruling our deep-seated natural emotions is never easy. Freud knew that well enough, but he was wrong in his speculations that "primitive man was better off knowing no restrictions of instinct." As Wright points out,[13] this is a mere legend. It has been a long, long time since any "primitive man" could enjoy "no restrictions" on these "instincts." Repression and the unconscious are the products of evolution too, and were well developed long before civilisation further complicated human mental life.

During the long Mesozoic period, when the daylight hours were dominated by predatory reptiles, the members of the early mammal

---

13   Wright, *The Moral Animal*, 323.

lineages kept out of their way by adapting to life as small secretive animals active mainly at night. They swapped the keen colour vision they had inherited from fish, their last common ancestors with reptiles, for characteristics more suited to nocturnal life, such as acute hearing, warm blood, and fur. Humans have more recently recovered the advantages of colour vision, but still share the additional features, such as night vision, good hearing, and strong emotions with other mammals, such as dogs.

The Forebrain

The well-known curly cover wrapped right across the top of the total structure, the forebrain or neocortex, is by far the largest part of the human brain. It has developed so strongly in us that it has changed the shape of our skull, adding a large rounded lump on the top. It is the seat of consciousness, language, and thinking, and its job is to weigh up the information coming from the lower centres, analyse options, and make rational decisions between conflicting stimuli. It is aware of the irrational *biological* urges sent up powerfully from the hind- and midbrain, but cannot totally silence them. It is never immune to thoughts such as ("aaaaah, that is an attractive body, I *want* to get close to it") versus the rational, *social* imperatives and options-weighing facility stored in the neocortex ("impossible, the boss is watching"). Multiple recent studies in primatology show that we share this capacity with our closest mammalian cousins, the primates.

The frontal lobes of the forebrain are the location of the human capacity to develop a higher purpose. This part of our brain is unique to humans. From here we can survey the human endeavour in its broadest terms, and perceive the significance of matters beyond our individual interests. Here is where we decide on, or avoid, the self-discipline needed to commit ourselves to purposes other than our own. Here, if anywhere, we learn to control our inner conflicts of interests and practice the virtues of moral choices and community engagement. As Michael Dowd points out,

Understanding the unwanted drives within us as having served our ancestors for millions of years is far more empowering than imagining that we are the way we are because of inner demons, or because the world's first woman and man ate a forbidden apple a few thousand years ago. The path to freedom lies in appreciating one's instincts, while taking steps to channel these powerful energies in ways that will serve our higher purpose.[14]

The Origins of Moral Dilemmas

These inevitable inner conflicts are the stuff of all our experiences of interpersonal dilemmas. Most importantly, they are not the result of anything that might or might not have happened in some hypothetical garden during the Iron Age, but of the structure of our brains evolved over millions of years of vertebrate evolution. The ancient concept of original sin has value in identifying our inner predisposition to self-centred actions, but the conclusion that any human ancestors were responsible has not.

Augustine's proposal that human nature is fatally flawed, together with the related idea that sole power of forgiveness should be reserved to the institutional church of the west, was based on politics, not theology. According to theologian Elaine Pagels, one of several reasons why Augustine's theory of the Fall eventually triumphed was that it made palatable the uneasy alliance between the Catholic church and Roman imperial power.[15] Augustine had many opponents, but his theory had a vital competitive edge at a time when the most pressing question was the urgent need to make sense out of the new interdependence of church and state.[16]

The Roman Catholic theology of the Fall is not only contradictory to human nature, it is also completely at odds with both earlier rabbinic and with later Eastern Orthodox traditions. Nevertheless, it was followed throughout the western world until the Enlightenment made

14   Dowd, *Thank God for Evolution*, 162.
15   E. Pagels, *Adam, Eve, and the Serpent* (London: Penguin, 1988), 126.
16   A. Kee, *Constantine versus Christ: The Triumph of Ideology* (London: SCM Press, 1982).

the collision with Augustine's teaching a central, tragic plank of the unnecessary war between science and religion. It has taken centuries of trying and rejecting alternative interpretations to reach the one that seems rational to us today. As Polkinghorne explains, we live in "an evolutionary world to be understood theologically as a world allowed by the Creator to make itself ... The picture is of a world endowed with fruitfulness, guided by its Creator, but allowed an ability to realise its fruitfulness in its own particular ways."[17]

The Evolution of the Human Brain

The question arises, why is our brain constructed in this complicated form? The answer can be best illustrated by revisiting the ancient advice against reinventing the wheel. A wheel is a modular unit first invented in ancient times to reduce the effort needed to drag a heavy load along the ground. Its capacity to minimise friction was later used in hundreds of other contexts, from chariots to wrist watches. The early lorries added a new idea, an engine, to better advance the capacity of load-bearing vehicles.

The same idea explains the origins of the human brain. The ancient anatomists recognised the three-part structure, with the whole gradually becoming larger in the higher animals, but they believed all creatures were created separately. So animals from fish to humans were arranged in a natural scale of independent rungs on a ladder from simple to complex, with later abilities eclipsing earlier ones. Now we can agree that the brains of all creatures have the same three parts going right back to the early fish, 450 million years ago, where the earliest versions of three parts are visible as modest bumps at the head end of the spinal cord.

The three parts had the great advantage of being modular units, that is, capable of being added to and modified in the course of evolutionary history. The hindbrain's job has not changed much since it

---

17 J. Polkinghorne, *Quarks, Chaos and Christianity: Questions in Science and Religion* (London: Triangle SPCK, 1994), 42–43.

was inherited by the early reptiles in the Mesozoic era, which started about 250 million years ago. Its integrated control of basic metabolic functions, plus some reptile-specific additions, has simply been copied down every generation en bloc, which is why we can describe our hindbrain as our legacy from the reptiles. Any modern textbook of evolutionary zoology will include diagrams illustrating the long process of development, deduced from the fossil record.

In the early mammals, starting in about 160 million years ago, the midbrain developed emotional capacities not known to reptiles, and they became added to our lineage. In time, the early hominids of about 2 million years ago inherited all that their ancestors had had, and also hugely expanded the neocortex. Finally, the first true humans refined the frontal lobes, the thinking part that makes us truly human, along with the origin of language only about 200,000 years ago.[18]

Over the last couple of hundred years we have learned much more about our ancestry from palaeontology, anatomy, neurophysiology, and genetics. The story becomes more comprehensive, and yet more marvellous, with every new discovery.

The Origins of Morality

Contrary to earlier assumptions, morality is far from being a cultural imposition unique to humans, although in us the cultural dimension is dominant. There is a substantial case for the view that evolutionary processes be accepted as part of any contemporary theory of morality.

> Moral reasoning is done by us, not by natural selection. But at the same time ... human morality cannot be infinitely flexible ... Natural tendencies may not amount to moral imperatives, but they do figure in our decision-making.[19]

---

18   R. Byrne, *The Thinking Ape: Evolutionary Origins of Intelligence* (Oxford University Press, 1995), 162.
19   F. van de Waal, *Good Natured: The Origins of Right and Wrong in Humans and Other Animals* (Cambridge, MA: Harvard University Press, 1996), 39.

All the same, the apparently counter-intuitive transition from animal-level evolutionary egoism ("Look after Number One") to true human ethics still requires explanation.

The most likely explanation of the development of true ethics is that this is another example of the way natural selection can modify a character evolved for one purpose and adapt it to serve another. Whales' flippers and bats' wings are analogous with reptilian feet, and mammalian ear bones are derived from fish jaw bones, simply because evolution is a cumulative process, and the material available for new forms is determined by what has survived from previous forms. Animals are necessarily compromises of design,[20] and their ability to take advantage of the opportunities opened up for them by environmental change is constrained by the history of their lineage and by existing genetic variability. The process works as well on behavioural traits as on feet and wings.

The Advantages of Intelligence

One of the most convincing explanations for the evolution of intelligence is that it allows more scope for social manipulations leading to sexual advantages within a group. These require recognition of individuals, and memory for previous transactions with known group members. Life in a primate group demands skill in navigating the continually shifting alliances that determine personal status and breeding success. Brainier chimps are simply better players of games of repeated exchanges of favours, leading to more mating opportunities. Greater skill in this is certainly rewarded; for example, the alpha male of a band of chimps is not necessarily the strongest one, but the one best able to maintain a dominant position by the manipulation of alliances with others.[21] Once evolved and further refined, as in modern humans,

---

20 N. Eldredge, *Reinventing Darwin: The Great Evolutionary Debate* (London: Weidenfeld and Nicolson, 1995), 46; G. C. Williams, *Plan and Purpose in Nature* (London: Weidenfeld & Nicolson, 1996).
21 Byrne, *The Thinking Ape*, 195–200.

intelligence was available to be applied to cultural skills, such as abstract mathematics, astronomy and music.

Similarly, the emotions that evolved to assist groups to maintain their cohesion by reciprocal altruism were available to be extended to what Waal[22] calls genuine community concern among chimpanzees. It is not, Waal is careful to point out, that these animals worry about the community as an abstract entity, more that they prefer to maintain the kind of peaceful, cooperative community that is in each of their own best interests. In evolutionary terms it is a short step from there to systems of conscious ethical rules.

Once evolved for related but different purposes, community concern allied with reflective intelligence became available to be refined into genuine, selfless altruism characteristic of the real spiritual world. In turn, each of these characters enhanced the individual breeding success of our far distant ancestors. With time and sociality (i.e., repeated encounters with the same individuals), the ruthless computations of competing self-interest pass from "Me first" to "Cooperation pays." Egoism in the primates has passed from "Look after Number One" to "Scratch my back and I'll scratch yours." Human ethics have thereby grown beyond the original dependence on animal precursors.

Given that background, we can begin to formulate a very different and much less destructive view of what has always been labelled as human immorality, and especially our supposedly inbuilt selfishness, long labelled original sin.

The Eden Myth for Space-Age Kids

Just because traditional myths are embedded in language no longer acceptable today does not make their ancient truths no longer true. One powerful way to defuse the war between modern science and ancient religion is to rediscover the wisdom of our ancestors by recasting their traditional myths into new stories conveying the same truths in con-

---

22   Waal, *Good Natured*, 205, 117.

temporary terms, more appealing to modern imaginations, especially of children.

For example, the biblical account of the conversation around the apple tree in Eden sounds entirely ridiculous if read literally (one parent was quoted on social media as angrily demanding that no one should teach his children any nonsense about "talking snakes"). Its definition of sin also sounds absurd in light of modern rules limiting judicial proceedings to the guilty parties, not to their descendants. But the same story can appear quite different if retold in terms of an imaginary conversation between the conflicting parts of the human brain, even when we retain the exact words of the original texts.

The Voice of Our Hindbrain

Put aside for a moment any distracting doubts about the reality of talking snakes, and remember that the real, documented, and active reptilian legacy within our own brains *in the here and now* is perfectly represented in the serpent of Eden. It was, yes, a reptile. The conversation can be reimagined into new terms unknown to the authors of Genesis but in their own words, as follows.

> Now the serpent was more crafty than any of the wild animals the Lord God had made. He said to the woman, "Did God really say, 'You must not eat from any tree in the garden?'" The woman said to the serpent, "We may eat fruit from the trees in the garden, but God did say, 'You must not eat fruit from the tree that is in the middle of the garden, and you must not touch it, or you will die.'" "You will not certainly die," the serpent said to the woman. "For God knows that when you eat from it your eyes will be opened, and you will be like God, knowing good and evil" (Genesis 3:1–5).

Eve's hindbrain's suggestions can sound all too familiar to anyone questioning an authoritative but apparently illogical prohibition.

The Voice of Our Midbrain

The seat of our heedless emotions and ambitions prompted Eve to greatly desire what the serpent had promised, but didn't warn her to stop to think of the consequences:

> When the woman saw that the fruit of the tree was good for food and pleasing to the eye, and also desirable for gaining wisdom, she took some and ate it. She also gave some to her husband, who was with her, and he ate it (Genesis 3:6).

The Voice of Our Forebrain

Suddenly confronted by higher authority, caught red-handed and urgently surveying its options, Adam's forebrain realised its danger and tried to find a way to avoid being held responsible for imminent disaster. It was a classic piece of buck-passing, easily recognisable today. The man blamed not only the woman, but God himself for providing such an unsuitable companion, whereupon the woman blamed the serpent:

> The man said, 'The woman you put here with me—she gave me some fruit from the tree, and I ate it.' ... The woman said, 'The serpent deceived me, and I ate' (Genesis 3:12–13).

The Voice of Our Frontal Lobes

This, the truly human part of us, is the only part capable of seeing a higher purpose and a survival tactic even in the aftermath of tragedy: "Adam named his wife Eve, because she would become the mother of all the living" (Genesis 3:20).

When operating together as a disciplined unit, and always kept under control by well-developed frontal lobes, the various components of a human brain can create a fully human mind ready to teach its owner to grow into a mature member of rational, civilised humanity. On the other hand, a brain with only poorly developed higher functions is less able to avoid the sort of behaviour powerfully prompted by its baser instincts. The personal and social consequences of such incomplete development certainly parallel what conservative theology has always labelled as sinful. But a more informed and compassionate view based on science can remove the burden of ages-worth of guilt and grief.

Surely, no better reason could be found to integrate the insights of science and religion.

Genesis 1

Of the two creation stories in Genesis, the one that has caused the most strident disputes between science and religion is the first presented, although written much later. By contrast with the ancient oral tradition preserved in Genesis 2-3 concerned with all humanity, Genesis 1 is a literary work dated to around 550 BCE and later, written in the style of the Priestly circle of Jerusalem. They and their cultic interests became prominent during the Exile starting in 587 BCE, when most of the population of Israel was deported to Babylon (2 Kings 24) and Solomon's Temple was destroyed (2 Kings 25).

The concern of the Priestly authors was focused on the people of Israel. Their version of the creation story does at least introduce the creatures in roughly the right order, by our standards—vegetation before birds and fish, and land animals before humans. Despite this passing superficial resemblance, nothing prevented certain Christians from using Genesis 1 to contradict science. This is a category error of the worst kind, understandably stimulating multiple defence strategies from both sides. The situation is a perfect trap for the uninformed enthusiasts, each equally outraged by the others' misinterpretation of their own position.

Ironically, much of their endless futile argument could have been muted if the participants had taken more notice of one of St Augustine's lesser known works, entitled *The Literal Meaning of Genesis*. Pointing out that non-Christians already know something of the science of their day, Augustine warns that "It is a disgraceful and dangerous thing for an infidel to hear a Christian, presumably giving the meaning of Holy Scripture, talking nonsense on these topics; and we should take all means to prevent such an embarrassing situation."[23]

The Long Shadow of Ancient Cosmology

Faced with new information, we all search for an explanation that fits with what we already believe, whether or not our idea is what the author intended. When it comes to understanding a part of our world that is too small or too large to be seen with our own eyes, we have to construct a model of it.[24] Misinterpretation of models expressed in authoritative written words is especially easy. The science-religion conflict is too often based on centuries of imposing our own cultural assumptions upon an ancient text, and failing to ask the right questions on what it was originally about.

We live in a materialist culture, and our assumptions of how the universe works (the subject of modern cosmology) colours our thinking in ways we seldom recognise, and which was certainly completely unknown to the authors of Genesis.[25] We leap to the conclusion that Genesis 1 is describing the origin of the material universe, because we can't see how else it could be read. We assume that the obvious contradiction between Genesis 1 and evolutionary science arises because the biblical writers were ignorant of science, and their story can be dismissed as a fable. But those who take the trouble to understand how ancient cultures thought about their world tell us that the real primary concern of Genesis 1 was quite different. Hebrew theologians did not ask, "How was the world made?" But "What is it *for*?"[26] We misread the

---

[23] Saint Augustine, *The Literal Meaning of Genesis*, trans. J. H. Taylor (New York: Newman Press, 1982), 42–43.
[24] C. M. King, "Models of Invisible Realities: The Common Thread in Science and Theology," in *Creation and Complexity: Interdisciplinary Issues in Science and Religion*, ed. C. Ledger and S. Pickard (Adelaide: Australian Theological Forum, 2004), 17–48.
[25] G. J. Glover, *Beyond the Firmament: Understanding Science and the Theology of Creation* (Chesapeake, VA: Watertree Press LLC, 2007).
[26] Walton, *The Lost World of Genesis One*, 26.

whole story because we fail to understand this absolutely crucial difference between our world and theirs.

Genesis 1 Is about the Sovereignty of God, not the Origins of Life

The most accessible and recent guide to help us understand this primary text is John Walton's book, *The Lost World of Genesis One*. It shows how, when we learn to ask the right questions about the original meaning of Genesis 1, any reasonable grounds for the dispute with science disappear altogether. Genesis 1 does not contradict science—it is not *about* science. On the contrary, it is concerned only to assert the Hebrew belief in the authority of God overruling all the ancient cosmologies common to all cultures of 3000 years ago. Pagans saw the universe as created by multiple deities for their own pleasures, and the human population as living in slavery and fear of them. Contrary to that, Genesis 1 is a masterly statement of the Hebrew belief in a world created by one, all-powerful, and loving deity, specifically for the benefit of human creatures capable of enjoying and caring for it.

The logic is very clear when the six days are arranged in two columns of three. Reading down the columns from days 1–3 shows the creation of functional spaces in order. They provide the bases of time (day and night), weather (water and sky), and food (land and vegetation). Reading across the rows shows the sequence of insertion of inhabitants into the functional spaces made ready for them. On day 4, the sun, moon, and stars appear, responsible only to provide the visible markers of time, not light itself. On day 5, the waters and the sky are inhabited by fish and birds, and commanded to fill the earth. On day 6, the land and vegetation are occupied by beasts, whose function is to serve humans, and people, who in turn are responsible for caring for the earth and its inhabitants.

In short, the text insists that the sun, moon, and stars are creatures, not gods, and are certainly not to be worshipped. The dome of the sky was seen as a solid firmament, with windows to let through the rain, and fixed tracks along which the sun, moon, and planets moved. The heavens and their inhabitants were created to serve humanity, by marking the passing of the days and seasons, and helping us to organise the annual rounds of planting and harvesting. The basic assumption was that things exist, not because they have definable material properties, but because they have a *function* in an ordered system.

Walton draws an illuminating analogy between the sequence of divine actions in Genesis 1, and the building of a new school as summarised in six stages. First, the designers have to set out the structure, so on day 1 they need light on their plans. On days 2 and 3 they need to build all the required functional spaces (classrooms, library, gym, offices, gardens, playing fields, pool). Only when these are ready can those spaces be populated with inhabitants; on day 4, electric lights, power points, clocks, and internet; on day 5, aviaries, terraria for frogs and lizards, aquaria, and fishponds; and finally on day 6, pupils and staff.

More significantly, our materialist world view does not prepare us to appreciate the vital importance of the seventh day. Genesis says that God "rested on the seventh day from all his work which he had done." We get the idea of God being tired out, and sitting back with his feet up. The original readers would have understood the words quite differently.

When Eastern cosmologies talked of their deities "resting" in their own temples, they meant that they took up residence there. So Genesis is declaring that on the seventh day the whole cosmos became God's temple, his residence from where he continued his work of upholding all creation. Hence, in 970 BCE Solomon built a temple for the most visible sign of God's presence, the Ark of Covenant (1 Kings 8:4). The critical point to grasp from Walton's book is that the function of the cosmos is to provide the residence of God, and, since function is the prerequisite for existence, without that function the cosmos would not exist.

Genesis 1 as the Text of a Communal Festival Celebrating the Cosmos

Read with understanding of its original intent, Genesis 1 does not contradict science at all. On the contrary, it is set out as a text for congregational participation in a joyful annual festival. It has rhythmic wording suitable for group speaking; it has a strong emphasis on the world designed as home for people; and is regularly punctuated by choruses proclaiming that "it was good." In this context, "good" means fit for purpose, not morally good. For example, the arrival of Eve was good because the human condition is not functionally complete without both genders.

Walton argues cogently that all these features make the most likely original context of Genesis 1 as providing the text for a regular reenactment of a literal seven-day festival.[27] Important events, like the inauguration of Solomon's temple, were often celebrated in public festivals running for several days, as described in 2 Chronicles 7:8. The inauguration ceremony for the cosmic temple, celebrating the Hebrew vision of the functional origins of the cosmos, would certainly deserve a really special annual ritual.

Genesis 1 does not describe the material origins of the earth, because everything was simply assumed to have been made by God. The questions we ask of the text, such as, how could there have been light on the first day when the sun did not appear until the fourth day, would have been pointless and incomprehensible to those for whom it was written. Walton's analysis shows that Genesis 1 is not and never was intended to explain the material origins of the universe in terms that have any relevance to our scientific knowledge. Only much later did philosophers begin to suspect there could be more to see behind the solid firmament of the sky.

It takes a deliberate effort for us to cast off our materialist assumptions and step out of our world into that of 3000 BCE. But if we do, we discover that Genesis 1 does *not* require us to choose between

---

27 Walton, *The Lost World of Genesis One*, 90–162.

loyalty to an ancient religious belief versus intellectually acceptable contemporary science for explanations of the world around us. More importantly, it does *not* deny a religious assertion that God made material creation, it says only that Genesis 1 is not about that story.

The critical point to grasp is that Genesis 1 does not deny evolution, or that the material universe evolved long before humans; rather, it assumes that the long procession of prehuman creatures appearing on Day 5 helped to prepare the earth for beasts and humans. (Yes, evolutionary science can confirm that fish and birds appeared on earth long before modern mammals and humans.) They were like the necessary rehearsals before the performance of a play, but the rehearsals are not the play, says Walton.[28] Rather, the cosmic play finds its meaning when the audience is present, because the play exists for them. Science can find meanings too, but different ones. Since Genesis 1 never was about material origins, there is no conflict with science. The science-religion war was *never necessary*. On the contrary, science owes much to the Hebrew (not literalist) theology of creation.

Historic Creationism Is the Most Ancient Basis of Science

Unlike most other cultures of their time, the Hebrews insisted that trees, rivers, and rocks did not have their own resident spirits, but that all matter was merely matter, open to human use and investigation. Western technology has inherited this attitude, and is therefore seen to have been responsible for a systematic, historic campaign to demythologise nature. One unfortunate consequence is that any protection that superstition had once afforded the natural world was removed, opening the way to the unrestrained exploitation that has produced the modern ecological crisis.[29] Yet that very same demythologising doctrine also laid the foundations of modern science.

The Hebrews' understanding of the natural world was not "scientific" in any respect; they had no concept of "nature" as a separate en-

28 Walton, *The Lost World of Genesis One*, 97.
29 L. White, "The Historic Roots of Our Ecologic Crisis," *Science* 155:3767 (1967): 1203–1207.

tity. But they assumed the total obedience of nature to the universal rational laws laid down by a rational creator. More importantly, humans are also rational, therefore confidence in human rationality allowed us the intellectual freedom to explore the world, free of all the old fears of retribution from angry pagan deities. The same assumptions were taken up by the Arab astronomers and mathematicians who contributed so much to the science of non-Christian cultures and the preservation of ancient Greek philosophy during the Middle Ages.[30]

The three main themes of the historic creationist tradition assert that the universe reflects the goodness, rationality and freedom of God, and therefore creation itself must be good, rational, and contingent. These assumptions were in due course incorporated within Christian faith. Christianity was therefore open to science from the beginning, and this indeed is one of several reasons why the roots of modern science are deepest in the Christian west.[31] But that is only part of the story.

Modern science also owes much to early-modern Renaissance and to medieval philosophies of nature, which were strongly influenced by Arabic natural philosophy derived at least in part from Greek, Egyptian, Indian, Persian, and Chinese texts. These rested, in turn, on the wisdom generated by other, still earlier cultures. One historian has called this twisting braid of lineage "the dialogue of civilizations in the birth of modern science."[32] Recognising that modern science grew out of the give-and-take among many cultures over centuries does not disparage the crucial role of early- modern Protestants and Catholics in casting the moulds within which modern science grew. But the Christian vision contributed much to the rich diversity of the cultural and intellectual soil into which the roots of science extend.

---

30   J. H. Brooke, "Contributions from the History of Science and Religion," in *The Oxford Handbook of Religion and Science*, ed. P. Clayton and Z. Simpson (Oxford: Oxford University Press, 2006), 293–310.
31   I. G. Barbour, *Religion and Science: Historical and Contemporary Issues* (New York: HarperCollins, 1997), 28.
32   N. J. Efron, "Myth 9: That Christianity Gave Birth to Modern Science," in *Galileo Goes to Jail and Other Myths About Science and Religion*, ed. R. L. Numbers (Harvard University Press, 2009), 79–89.

The three concepts of goodness, rationality, and contingency are all vital for science. If the universe is functionally good, it is worthy of careful study; if it is rational, it is predictable and reliable; and if it is contingent it could have been otherwise than it is, so the state of things has to be studied by experiment, not deduced from pure reasoning. Moreover, the ancient tradition insisted that there had to be a fruitful balance between the rationality and the freedom of God in creation: if rationality is overemphasised, the universe becomes fixed and uninteresting, whereas if freedom is overemphasised, the universe becomes incoherent, unpredictable, and impossible to study. In a nutshell, if the world is not rational, science is not possible; if the world is not contingent, science is not necessary. Thus, the historical relationship between theology and science in the western world has been very much more long-standing, complex, productive, and positive than many participants in the present debate may realise.

On the other hand, Christianity should not, and does not need to, defend itself by claiming credit for having contributed to the rise of science, which would expose it to the developing contemporary backlash against the excesses of scientific technology. The most it need claim is that true Christianity is not, and never has been, incompatible with true science.[33] C. S. Lewis neatly illustrated this compatibility when he put into Screwtape's mouth the advice (to a young devil attempting to ensnare an unsuspecting human soul),

> Above all, do not attempt to use science (I mean, the real sciences) as a defence against Christianity. They will positively encourage him to think about realities he can't touch and see. There have been sad cases among the modern physicists. If he must dabble in science, keep him on economics and sociology.[34]

---

33 A. Peacocke, *Theology for a Scientific Age*, enlarged edn (London: SCM Press 1993), 76.
34 C. S. Lewis, *The Screwtape Letters* (London: Geoffrey Bles, 1942), 14.

Sound advice indeed—and in view of the diabolical consequences of modern market economics, one might deduce that Screwtape's pupil has been remarkably successful in following it.

Layers of Explanation

The simplistic use of Genesis to set science against religion or vice versa falls into an ancient intellectual error, invisible to most modern writers unfamiliar with the logic of inference. They do not see the dangers of imposing their own one-dimensional cultural assumptions upon a classic text originally conveying a quite different message. As John Haught explains:

> Everything in our experience can be explained at multiple layers of understanding, in distinct and non-competing ways ... [This idea] is an ancient one, endorsed by Socrates, Plato, Aristotle, Augustine, Aquinas, Kant, and many other great thinkers ... a page of a book exists because a printing press stamped letters in black ink on white paper ... [and] because an author is trying to get some ideas across to his readers.. [and] because a publisher [published it]. These are not competing explanations.[35]

> [Dawkins] keeps asking, where is the evidence—and here he clearly means scientifically available evidence—of any divine principle of meaning and directionality in life ... [But] meaning and purpose cannot show up at the level of scientific analysis. As far as he is concerned, science is powerful enough in its intellectual sweep to answer every conceivable question about the natural world. But this is a belief ... that demands from science a kind of insight that it cannot in principle provide ... Layers of causality are not mutually exclusive ... [The chemistry of printing tells us nothing about the author's intention] ... The rules of grammar are essential, but meaning is not determined by them.[36]

---

35   J. F. Haught, *Making Sense of Evolution* (Louisville, KY: Westminster John Knox Press, 2010), 23.
36   Haught, *Making Sense of Evolution*, 70–71.

In some respects, the distinction between the two creation stories in Genesis parallels the idea of the evolution of religion as proposed by Fraser Watts.[37] Watts suggests that Genesis 2–3 represents the older, intuitive, oral tradition, to which was later *added* the more conceptual, rational doctrine propounded by the Priestly authors of Genesis 1. Biologists see a comparable additive process in the physical evolution of brain functions, whereby the new capabilities of the mammalian brain have been built up on the original basic structure inherited ultimately from Devonian fish. There is therefore no contradiction between evolutionary biology and Watts (or Robin Dunbar, who suggested a similar distinction), that this evolutionary process is a matter of *adding to* earlier religious insights, *not* replacing them with later ones.

Consequences for Science Education

The net result is double jeopardy for our young people. Like all of us, they search for ideas that explain the world around them and give meaning to their personal lives. Students often reject the idea of evolution because they do not understand it, not because they understand it and find it wrong. I can confirm this from my own teaching experience. Likewise, unbelievers often reject the idea that Christianity could be rational or relevant to this age because they do not know there is any such thing as serious, critical theology, or because their view of what the church stands for has been coloured by the failings of its members.

If there really is no fundamental conflict between science and religion, we need to end this tragic and unnecessary situation as soon as possible. If our young people are to defend themselves against irrational beliefs bombarding their social media feeds daily, from both aggressive secularism and outdated preaching, they need to be equipped with a more realistic understanding of both science and faith. It *is* possible to do that: there are many thought-provoking articles and books on the in-

---

37   F. Watts, "The Evolution of Religious Cognition," *Archive for the Psychology of Religion* (2020): 42, 93, https://doi.org/10.1177/0084672420909479.

terface between science and religion, and even tertiary courses (some of them available on the internet) on the science-religion dialogue.

To paraphrase a well-known saying: All that is required for irrationality to triumph is that those who can think remain silent. Now is the time for thinkers to speak out.

The author reports there are no competing interests to declare.
Received: 16/03/22 Accepted: 09/08/22 Published: 13/08/22

Between Darwin and Dostoevsky: The Syntheses of Theodosius Dobzhansky

Christopher Howell

Abstract: Theodosius Dobzhansky was one of the foremost evolutionary biologists of the twentieth century who spent a great deal of time pondering, studying, and writing about religion. A confessed Eastern Orthodox Christian, though one with an idiosyncratic take on the faith, Dobzhansky was interested in harmonising the different elements of his life—religious background, scientific knowledge, and political beliefs. Throughout his oeuvre, he made various attempts to do this, and his legacy therefore amounts to a great synthesis. His greatest scientific achievement is the fusion of genetics and natural selection, which constitutes the groundwork for modern evolutionary biology. He also worked to synthesise democratic politics with Christian ethics, and religion with science. Dobzhansky was worried that science could not provide a basis for morality, and believed that Dostoevsky definitively proved this. Accordingly, he undertook not only to make sense of his own life and beliefs, but to protect and secure science, religion, morality, and democracy as parts of a cohesive whole.[1]

Keywords: Theodosius Dobzhansky; evolution; Eastern Orthodoxy; religion; science

---

Christopher Howell is an adjunct assistant professor of religious studies at Elon University. He holds a PhD in Religion from Duke University and a Masters in Theological Studies from Duke Divinity School.

1   This article incorporates material that first appeared as a series of essays on *Public Orthodoxy*, run by the Fordham Orthodox Christian Studies Center, and is reused with permission. The relevant pieces were published on 23 July 2021, 27 August 2021, and 13 January 2022. In its current form, the article develops new ideas and includes further material, presenting a consistent argument from beginning to end. The author is grateful to the *CPOSAT* reviewers for their useful comments and suggestions.

That one of the most important evolutionary scientists of the twentieth century was a confessed Orthodox Christian is an oft-overlooked and tantalising fact. Theodosius Dobzhansky (1900–1975), whose contribution to evolution and genetics was immense, remains an enigmatic figure in the history of science and religion. Philosophical questions of ethics, politics, and religion occupied him throughout his life, but his idiosyncratic religious ideas have not usually been probed as much as his scientific contributions.[2] This is understandable, of course, as there is no doubt that his legacy in the sciences is as secure as his legacy in religion is obscure. Nevertheless, it is worth analysing his scientific and religious beliefs alongside each other and in depth, as they most certainly influenced each other. As Jitse M. van der Meer argues, Dobzhansky was driven by a desire to harmonise Darwin with his Eastern Orthodox background—this quest implicitly drove his scientific research program.[3]

When studying Dobzhansky's thought in detail, it becomes clear that the quest for *synthesis* was the dominant intellectual thrust behind his philosophical excursions. He hoped to find ways to integrate his scientific knowledge with his religious life, to bridge what has so often been torn asunder. As he wrote towards the end of his life, the Delphic command to "know thyself" extends beyond science, but science must be included. "This adds up to something pretty simple," he observed, "a coherent credo can neither be derived from science nor arrived at without science."[4] Dobzhansky came to evolution through philosophical interest, as Garland Allen notes, and so it is not surprising he maintained

---

[2] There are a few noteworthy exceptions. Michael Ruse addresses Dobzhansky's religion and philosophy in-depth in his chapter "Dobzhansky and the Problem of Progress" in the volume *The Evolution of Theodosius Dobzhansky*, ed. Mark Adams (Princeton, NJ: Princeton University Press, 1994), 233–245. Jitse M. van der Meer cites Ruse as the main scholar who engaged in such analysis, other than van der Meer's own work on the subject. See Jitse M. van der Meer, "Theodosius Dobzhansky: Nothing in Evolution Makes Sense Except in the Light of Religion," in *Eminent Lives in Twentieth-Century Science and Religion*, ed. Nicolaas Rupke (Bern: Peter Lang, 2009), 105–127.

[3] Van der Meer, "Theodosius Dobzhansky," 113–116.

[4] Theodosius Dobzhansky, *The Biology of Ultimate Concern* (New York: The New American Library, 1967), 9.

an interest in philosophy and religion throughout his life.[5] Charles E. Taylor relates that Charles Birch, who wrote on philosophy despite being a scientist, decidedly influenced him; it is Birch who inspired Dobzhansky to do the same.[6] To build a holistic worldview where science, religion, and philosophy hold together, Dobzhansky embarked upon a lifelong journey. This drive was about more than his personal interest, however, as he was concerned that a purely scientific picture of reality might not be able to account for ethical principles like human equality, which he viewed as the basis for democracy.

In this paper, I will consider three of Dobzhansky's syntheses. In his desire to heal fractures in human knowledge and experience, he bequeathed three important attempts—to synthesise natural selection and genetics, democracy and ethics, and religion and science. The first is what brought Dobzhansky his fame. This "modern synthesis" is well known, and therefore much of the material discussed here is established already in the secondary literature. Consequently, this portion will be something of an overview of scholarship. However, little attention has been given to Dobzhansky's political views, while a little more (but still not enough) has been said about his religious beliefs. In addition, how the latter vouchsafed the former remains a poorly researched topic. The second and third parts of this paper will focus on these elements and put his neglected philosophical books, now out of print, in conversation with his science. I will attempt to prove that the synthetic approach that defined his science—for which he was famous—extended into politics and religion as well. Dobzhansky spent considerable intellectual energy bridging the gaps between these areas of human experience and tying them together into a holistic framework. His legacy, then, is that of a great synthesiser.

---

5 Garland E. Allen, "Theodosius Dobzhansky, the Morgan Lab, and the Breakdown of the Naturalist/Experimentalist Dichotomy, 1927-1947" in *The Evolution of Theodosius Dobzhansky*, 94.
6 Charles E. Taylor, "Dobzhansky, Artificial Life, and the 'Larger Questions' of Evolution," in *The Evolution of Theodosius Dobzhansky*, 165–166.

Dobzhansky's Life and Background

The unusual name of the Ukrainian-born Theodosius Dobzhansky was consequence of his mother's prayers. As recounted by his daughter Sophia, "My father's parents were childless for quite a while after their marriage and tried to remedy their condition by prayer and pilgrimage." Their prayerful journey took the couple to the shrine of St Theodosius of Chernigov, and when they found themselves with child, they christened him with the saint's name. Dobzhansky was thus enmeshed in Orthodox religious culture from his birth—though, interestingly, many of his paternal ancestors were Polish Catholics who converted to Orthodoxy in the late nineteenth century. On his mother's side, Dobzhansky was descended from a long line of priests, and his affinity for Dostoevsky was as much genetic as aesthetic, for he proudly numbered the great novelist among his maternal ancestors as well.[7]

When he was young and Russia was in the throes of revolution, Dobzhansky felt the "urgency of finding a meaning of life ... in the bloody tumult." But he was stuck between two poles that drew him equally: religion and science. He loved Darwin and he loved Dostoevsky. "The intellectual stimulation derived from the works of Darwin and other evolutionists was pitted against that arising from reading Dostoevsky," he wrote towards the end of his life.[8] Resolving this tension—which partly stands for the broader tension between his scientific interests and his religious background—became one of the driving forces of his career. When looked at more deeply, though, there was one particular struggle that occupied him. Darwin had unlocked the key to evolution, but Dobzhansky believed that Darwin—and scientific worldviews based on his thought—provided no real basis for ethics, especially the ethics of human equality. Furthermore, he felt that Dostoevsky had articulated the terrible truth of scientific atheism: that it has no ethics at all. He sought to find a way through this maze and preserve both science and religion in order to secure morality in both the personal and political

---

[7] Sophia Dobzhansky Coe, "Theodosius Dobzhansky: A Family Story," in *The Evolution of Theodosius Dobzhansky*, 13–14.
[8] Dobzhansky, *Ultimate Concern*, 1.

realms. In the coming decades, after he fled to America and became a "nonperson" in the USSR, Dobzhansky would emerge as one of the greatest biologists of the twentieth century. The search for union between the disparate spheres of his life continued to be dominant in all of his writing, however, not just his scientific research.

In America, his home from 1927 onwards, Dobzhansky's eccentricity made him memorable. Colleagues marvelled at his facility with languages (writing in fluent English despite only learning it as an adult) and were amused by his "extraordinary accent … high and staccato."[9] A scientist who joined him on one of his last field trips described him as "passionate and ready to take offence, but with a deep interest in the arts," and compared him to Vladimir Nabokov's unforgettable Timofey Pnin. This was a fitting comparison, as Nabokov followed Dobzhansky's scientific work with interest and the two corresponded in 1954.[10] In true Pninian fashion, Dobzhansky endured a "series of tragicomic rows with colleagues and officials that end[ed] up with his exile from New York and a forced move to the far west."[11] It was in California that Dobzhansky found a home and contributed his greatest scientific achievements—in between his favourite hobbies of mountain climbing in the Sierras and horseback riding in Pasadena.

The Modern Synthesis: Dobzhansky's Scientific Legacy

After shattering his knee in a horseback riding accident, Dobzhansky was bedridden and, in his own retelling, used the time to produce his most significant work: *Genetics and the Origin of Species* (1937).[12] This

---

9   E. B. Ford, "Theodosius Grigorievich Dobzhansky: 25 January 1900 – 18 December 1975," *Biographical Memoirs of Fellows of the Royal Society* 23 (1977): 60.
10 David M. Bethea, "Evolutionary Biology and 'Writing the Diaspora': The Cases of Theodosius Dobzhansky and Vladimir Nabokov," in *Redefining Russian Literary Diaspora* (1920–2020), ed. Maria Rubins (London: UCL Press, 2021), 144, https://doi.org/10.2307/j.ctv17ppc6w.10.
11 Steve Jones, "The day I went on a field trip with Theodosius Dobzhansky," *The Guardian*, 20 March 2016, https://www.theguardian.com/lifeandstyle/2016/mar/20/a-field-trip-with-theodosius-dobzhansky-steve-jones-genetics-biology.
12 William Provine was a little suspicious of Dobzhansky's memory, but nevertheless included his testimony of the events in a chapter on the man.

book proved pivotal for "the modern synthesis" of evolution, though its significance is lost now in the eighty-plus years since, when Darwinism went from being moribund to triumphant (in no small part due to Dobzhansky).

In the early twentieth century, evolutionary biology was in crisis, as the new science of genetics seemed to be incompatible with evolution by natural selection, Darwin's main contribution. Darwin did not know by what mechanisms heredity was transmitted, and he died before Gregor Mendel's pea plant experiments were rediscovered in 1900. But genetics was not easily integrated with evolution, at least not at first. William Bateson, who coined the word "gene" and popularised Gregor Mendel's ideas, doubted the harmony between genetics and the gradualism of natural selection. The famed geneticist Thomas Hunt Morgan likewise harboured some scepticism about Darwin's main theory, though he softened on this while Dobzhansky was a postdoctoral researcher at his Columbia University laboratory.[13] This period has come to be known as the "eclipse of Darwinism," in Julian Huxley's phrase. Darwin's theory of natural selection was diminishing, with many scientists preferring rival neo-Lamarckian theories such as orthogenesis.

Darwin was down, but not out. J. B. S. Haldane, R. A. Fisher, and Sewall Wright would construct the mathematical theory of population genetics, and Dobzhansky's *Genetics and the Origin of Species*, along with the work of Ernst Mayr and G. Ledyard Stebbins, would help build the edifice for the modern synthesis: the long-awaited marriage of natural selection and genetics. As Julian Huxley wrote, "The death of Darwinism has been proclaimed not only from the pulpit, but from the biological laboratory; but, as in the case of Mark Twain, the reports seem to have been greatly exaggerated, since to-day Darwinism is very much alive."[14]

See William B. Provine, "The Origin of Dobzhansky's *Genetics and the Origin of Species*," in *The Evolution of Theodosius Dobzhansky*, 99–114.

[13] Allen, "The Morgan Lab," 88; Nicholas W. Gillem, "Evolution by Jumps: Francis Galton and William Bateson and the Mechanism of Evolutionary Change," *Genetics* 159:4 (2001): 1383–1392, https://doi.org/10.1093/genetics/159.4.1383.

[14] Julian Huxley, *Evolution: The Modern Synthesis*, Definitive Edition (Cambridge, MA: MIT Press, 2010), 22.

In all this, Dobzhansky played the role of the synthesiser, translating the difficult mathematics of population genetics into readable language. As Peter Bowler writes, Dobzhansky "pointed the way toward a complete synthesis by presenting the mathematician's conclusions in a form [other scientists] could understand and use."[15] Dobzhansky's student Bruce Wallace agrees, writing, "It brought sense and logic to an otherwise completely muddled branch of biology."[16] It is hard now to even speak of evolutionary biology without using Dobzhansky's language. He brought into English the terms microevolution, macroevolution, gene pool, coadaptation, and homeostasis.[17] He helped develop the biological species concept.[18] Beyond that, in harmonising natural selection and genetics—which is an epochal achievement on its own—Dobzhansky concurrently helped merge the disparate scientific practices of naturalist fieldwork and experimental laboratory work. According to Garland Allen, in addition to genetics, "the more general fusion of the laboratory and field naturalist traditions ... remains among the deepest and most lasting aspects of Dobzhansky's legacy."[19] Scientific legacies are difficult, as works fall out of fashion in their respective fields quickly, but Dobzhansky's influence is clear. He even received the highest praise the ornery J. B. S. Haldane could give: Dobzhansky was good enough reason, and maybe the only reason, to visit America.[20]

A historical and biographical question is, then, why was it Dobzhansky that spearheaded this synthesis? Until recently, writing on Dobzhansky and his work tended to depict him as an American,

15   Peter J. Bowler, *Evolution: The History of an Idea* (Berkeley, CA: University of California Press, 2009), 336.
16   Bruce Wallace, "The Legacies of Theodosius Dobzhansky," in *Genetics of Natural Populations: The Continuing Importance of Theodosius Dobzhansky*, ed. Louis Levine (New York: Columbia University Press, 1995), 44.
17   Mark B. Adams, "Introduction: Theodosius Dobzhansky in Russia and America," in *The Evolution of Theodosius Dobzhansky*, 3; Wallace, "The Legacies of Theodosius Dobzhansky," 44.
18   Nikolai L. Krementsov, "Dobzhansky and Russian Entomology: The Origin of His Ideas on Species and Speciation," in *The Evolution of Theodosius Dobzhansky*, 31.
19   Allen, "The Morgan Lab," 87.
20   Costas B. Krimbas, "Resistance and Acceptance: Tracing Dobzhansky's Influence," in *Genetics of Natural Populations*, 23.

but, though he became a US citizen, to understand him one needs to *synthesise* both the Russian and the American aspects of his thought. This includes not only the Russian scientific tradition, such as Dobzhansky's debt to Yuri Filipchenko and Sergei Chertverikov, but also the philosophical and religious traditions.[21] Dostoevsky and Tolstoy are important, but so, too, is Vladimir Solovyov. Solovyov mediated much of Darwin's thought into Russia, where non-Darwinian evolution was less popular than in America. It is he who impressed on Dobzhansky the importance of progress and development in evolutionary history, a conviction that assisted him in sorting out the tangled relationship between natural selection and genetics. This led him, furthermore, to see evolution by natural selection as directional even though not "directed" (contrary to orthogenesis, which he viewed as deterministic).[22]

As Michael Ruse contends, it was Dobzhansky's religious views—influenced by Solovyov and others—that informed his scientific ones, such as his faith in developmental progress and his hostility to determinism.[23] Dobzhansky was vexed by the problem of evil, which might explain his affinity for Dostoevsky, and he believed Darwinian evolution allowed for free will, which would rescue the Creator from responsibility for extinctions. Wrote Dobzhansky, "predetermined [evolution] collides head-on with the ineluctable fact of the existence of evil ... the evolution of the universe must be conceived as having been in some sense a struggle for a gradual emergence of freedom." Darwin's theory meant that "the history of the living world has not been wasted."[24]

As Bowler speculates, Dobzhansky's fervour in defending a high anthropology and free will likely stemmed from his Orthodox roots.[25] But they were more than merely roots. While Dobzhansky's religious views were eccentric, they were real. Van der Meer chronicles that he tried to pray every morning and used Dostoevsky to bring his colleagues

---

21 Richard M. Burian, "Dobzhansky on Evolutionary Dynamics: Some Questions about His Russian Background," in *The Evolution of Theodosius Dobzhansky*, 138.
22 Van der Meer, "Theodosius Dobzhansky," 112.
23 Ruse, "Dobzhansky and the Problem of Progress," 239–240.
24 Dobzhansky, *Ultimate Concern*, 25, 120.
25 Bowler, *Evolution*, 345.

closer to God.²⁶ In turn, Costas Krimbas recalls that Dobzhansky insisted on making a pilgrimage to Mt Athos in order to take communion, but was evasive about why. He said it reminded him of childhood, but Krimbas surmised this was not the real reason.²⁷

Even though Dobzhansky's religious beliefs informed his science, they did not stay restricted to it. Rather, they would drive other attempts at synthesis—attempts to preserve democracy and to search for common grounds between religion and science.

Freedom and Equality: Dobzhansky's Political Views

It is in Dobzhansky's writing on ethics that Dostoevsky's influence, and the importance of religion to society, is most apparent. Freedom mattered to him. He was interested in articulating a scientific worldview where Darwin buttressed free will, and he felt Dostoevsky helped answer the problem of evil. At this juncture, Dobzhansky offered an early version of the "free process defence" to natural evil that anticipates John Polkinghorne's.²⁸ But he was also interested in protecting political freedom, both from totalitarianism and from hereditary aristocracy. His second synthesis amounted, then, to merging democracy with science and Christian ethics, to defend all three from conservative critics, whether of the religious, social, or economic bent. A hierarchical, aristocratic, class-based society was, in Dobzhansky's view, a defence mechanism designed to allay the fears of the wealthy when confronted with Jesus' harder sayings. "Christ's parable of the camel passing through the eye of a needle is too explicit to be easily interpreted away," he wrote. And he continued:

> To assuage their consciences, the Creator is blamed for having made some people nobles and others commoners, some wise

---

26 Van der Meer, "Theodosius Dobzhansky," 111.
27 Costas B. Krimbas, "The Evolutionary Worldview of Theodosius Dobzhansky," in *The Evolution of Theodosius Dobzhansky*, 188.
28 Van der Meer, "Theodosius Dobzhansky," 108; John Polkinghorne, *Belief in God in an Age of Science* (New Haven, CT: Yale University Press, 2003), 14.

and others improvident, some talented and others incompetent. Different people are thus born to occupy different stations in life. Such, allegedly, is God's will, and to go against it is sin.[29]

"Don't blame us," one can imagine the rich and the powerful saying, "it's God's fault for endowing us with superior genes." Wealth, power, influence, and so on, are simply inevitable under such circumstances, and no amount of political equality would change it.

Such hereditarians, observed Dobzhansky, were often political conservatives who believed "genetic conditioning of human capacities would justify the setting up of rigid class barriers and a hierarchical organisation of the society."[30] However, he argued, this was a misunderstanding of genetics and reflected a poor knowledge of inheritance. There is, he argued, no one-to-one relationship between genotype and phenotype, there is no "gene for" intelligence or any particular skill. Rather, genes allow for a "norm-of-reaction"—a pattern of phenotypic expression that flows from the genotype, but which can result in highly variable developments in each person as they grow, develop, and evolve. "A newborn infant is not a blank page," he wrote, "however, his genes do not seal his fate." The environment plays a crucial role.[31] Freud might have proclaimed that "biology is destiny," but Dobzhansky rejected this notion. "Heredity ... is destiny," he argued, "largely in proportion to our biological ignorance."[32]

Ironically, Dobzhansky argued, a rigid, caste-based society premised on stasis and a lack of change for the moneyed aristocracy would induce a great deal of genetic diversification at the top. Those with "superior" genes would easily beget offspring rather less like the *Übermensch* than they are wont to claim. In an ossified, isolated system, where natural selection could not operate, stagnation and devolu-

---

29 Theodosius Dobzhansky, *Mankind Evolving: The Evolution of the Human Species* (New Haven, CT: Yale University Press, 1962), 52.
30 Dobzhansky, *Mankind Evolving*, 247-248.
31 Dobzhansky, *Mankind Evolving*, 76.
32 Quoted in Diane B. Paul, "Dobzhansky in the 'Nature-Nurture' Debate," in *The Evolution of Theodosius Dobzhansky*, 223.

tion would be the name of the game. This was obvious to anyone who encountered the luxuriant upper class "snobs," self-styled elites, who were certainly "better endowed financially than genetically."[33] Dobzhansky was likewise contemptuous of any suggestions that there must be a social aristocracy of elite minds who stewarded culture and safeguarded it from the unwashed hordes. He singled out T. S. Eliot for criticism. "I, for one," he wrote, "do not lament the passing of social organizations that used the many as a manured soil in which to grow a few graceful flowers of refined culture."[34]

The solution to this was equality and its political expression, democracy. Inequality of opportunity prevents genetic change and allows for those ensconced at the top to maintain their wealth and status.[35] Equality, on the other hand, reduces "genetic wastage" and creates a more diverse society, beneficial to the entire species.[36] A static, changeless society—a non-democratic one—would in essence be conservative and unscientific. No wonder Dobzhansky highlighted that "the foundation of all conservatisms was undermined by the flood of scientific discovery."[37] In exchange, conservative hierarchical worldviews would naturally lead "to the frightful doctrines of Dostoevsky's Grand Inquisitor."[38]

In the end, Dobzhansky was a liberal with a tilt towards social democracy and a deep revulsion towards totalitarianism and hereditary authority. Despite the focus on democracy, however, he was suspicious of communism, which he termed a "Christian heresy." That he referred to famous communist works as "Marxist Scriptures" indicates that he viewed communism as a substitute religion.[39]

While he placed a high emphasis on human equality, Dobzhansky felt that it was an ethical precept and not one that could be reduced to a scientific postulate. This existentialist take on human dignity was likely influenced by his reading of Dostoevsky, and especially *The Broth-*

33 Dobzhansky, *Mankind Evolving*, 334.
34 Dobzhansky, *Mankind Evolving*, 325.
35 Dobzhansky, *Mankind Evolving*, 248.
36 Dobzhansky, *Mankind Evolving*, 324–325.
37 Dobzhansky, *Ultimate Concern*, 113.
38 Dobzhansky, *Ultimate Concern*, 106.
39 Dobzhansky, *Ultimate Concern*, 99; *Mankind Evolving*, 19.

ers Karamazov. "People do not need to be biologically (genotypically or phenotypically) alike to be equal before God," he argued.[40] Equality is, in essence, a Christian theological concept.[41] It is a good in and of itself, not because it may or may not be scientifically provable; good and evil, after all, are concepts beyond the capacity of science to articulate. Julian Huxley and C. H. Waddington may have laboured mightily to find an ethics based on evolution, but they failed. "The force of these strictures has never been overcome," contended Dobzhansky. Evolution by natural selection could, at most, "explain how we develop our belief that certain things are good and others evil; it does not explain why we *ought* to regard them good and evil."[42] In the end, no one could answer the Karamazovs. As Dmitri Karamazov summarises, "But what will become of people then ... without God and immortal life? All things are permitted then, they can do what they like?" The existentialists were right. Years later, Sartre famously captured the moral consequences of this belief: "Nor, on the other hand, if God does not exist, are we provided with any values or commands that could legitimise our behaviour. Thus we have neither behind us nor before us in a luminous realm of values any means of justification or excuse.—We are left alone, without excuse."[43]

But Dobzhansky couldn't leave it at that—his moral intuition was too strong. "Evil is," he wrote, "very real. Not only real but also unredeemable."[44] The reality of good and evil could not be explained scientifically because there is no gene for ethics. And ethics is needed because it presupposes the freedom necessary to safeguard democracy. "Attempts to discover a biological basis of ethics suffer from mechanistic oversimplification," he contended.[45] In turn, the knowledge of

---

40 Dobzhansky, *Mankind Evolving*, 52.
41 Dobzhansky, *Mankind Evolving*, 219.
42 Dobzhansky, *Mankind Evolving*, 343.
43 Jean-Paul Sartre, "Existentialism is a Humanism," in *Existentialism from Dostoevsky to Sartre*, ed. Walter Kaufmann (New York: Penguin, 1975), 353.
44 Dobzhansky, *Ultimate Concern*, 101.
45 Theodosius Dobzhansky, *The Biological Basis of Human Freedom* (New York: Columbia University Press, 1956), 131.

good and evil was given by revelation,[46] and we must remember that "the highest wisdom of all was at one time entrusted to a group of unlettered Galilean fishermen."[47]

All this points to Dobzhansky's hope to vouchsafe human equality, political freedom, and a society of open movement by grounding democracy in science and supporting it with Christian ethical concepts. Such was the second of his three syntheses. These multiple strands often seemed in tension, especially to his scientific colleagues, most of whom did not share his sympathy for religion. His first two syntheses, focusing on science and politics, were in fact conflicting: Darwin's world could not provide an answer to Dostoevsky's ethical challenge regarding a modern egalitarian society.

But could religion persist in a world of science? Dobzhansky believed there was difficulty in establishing a moral basis for human equality and democratic politics without religion. Accordingly, he hoped to achieve a third synthesis, one which would encapsulate, explain, and defend the other two: a harmony between science and religion.

Hope and the Ultimate Synthesis: Dobzhansky on Religion

Dobzhansky's religious views were idiosyncratic and highly personal. Charles E. Taylor lumps him in with "Russian Romanticism" and, while he considers his ideas interesting, nevertheless dismisses them as "outside analysis by reason."[48] Such a reductionist perspective need not prevent a deeper analysis of Dobzhansky's worldview, however. He considered himself Orthodox and so should be investigated with that kept perpetually in mind. Nevertheless, it must be admitted that the extent to which he held to specific Orthodox doctrines is unclear. Although he was open about his sympathy for religion and his interest in philosophy, he kept much to himself, praying in a language his colleagues could not understand. This has made his beliefs hard to parse.

---

46   Van der Meer, "Theodosius Dobzhansky," 111.
47   Dobzhansky, *Mankind Evolving*, 345.
48   Taylor, "Dobzhansky," 168

Ernst Mayr remarked that Dobzhansky believed in a personal God, and that "he would work as a scientist all week and then on Sunday get down on his knees and pray to God." However, Francisco Ayala, present with Dobzhansky when he died, maintained that he did not.[49] For his own part, Dobzhansky at times softened traditional dogmas, but he also wrote in *The Biology of Ultimate Concern* that it was "no use" to pray to a "deistic clockmaker God."[50] Yet Dobzhansky prayed often. How does one sort this out?

Belief is only one part of religious life. While Dobzhansky's beliefs were sometimes inscrutable, his practice was more overt. In his excellent essay on Dobzhansky's religion, van der Meer observes the way he was influenced by Solovyov but also includes a deep dive into Dobzhansky's diaries and journals to show that religion was a preoccupation throughout his life, not just as he approached death, as was sometimes thought. Dobzhansky did go to confession, although he did not appear to regard sin as significant as his colleagues would have expected—influenced as they were, even if they rejected it, by a more Protestant emphasis on depravity. As a consequence, he did not believe sin made it impossible to do good, maintaining his defence of human agency and freedom in the face of determinism (either scientific or theological). In fact, as van der Meer shows, the entries of Dobzhansky's diary were saturated with religion. He often began and ended with glorifications of God. He was bothered by the lack of religious education in America, writing that "the trouble is that they do not have moral and religious schooling, and that they grow up to be egoists and self-centered and freethinkers." He was disappointed with American Easter, penning a 1927 entry in his diary that could contend for the most Orthodox sentence ever constructed: "Easter is not interesting here; they buy special lilies or in general flowers and that is all. There is not even gourmet food, perhaps only two chocolate eggs. It has no

---

49   Michael Shermer and Frank J. Sulloway, "The Grand Old Man of Evolution: An Interview with the Evolutionary Biologist Ernst Mayr," *Skeptic* 8:1 (2000): 82; Francisco Ayala, "Theodosius Dobzhansky: January 25, 1900–December 18, 1975," *Biographical Memoirs of the National Academy of the Sciences* 55 (1985): 179.
50   Dobzhansky, *Ultimate Concern*, 98.

meaning."[51] Michael Ruse, likewise, contends that Dobzhansky's faith in God and hope for salvation was "nigh overwhelming."[52]

Hope was at the centre of Dobzhansky's religious worldview, and both Christianity and evolution offered it to him. Because evolution by natural selection allowed for a developmental process in history, and therefore made room for human freedom, it offered hope. As Dobzhansky stated in *Mankind Evolving*, the idea that humanity is not evolved but is, rather, evolving (much as, in Orthodox thought, humanity is not "once saved, always saved," but is, rather, always being saved), means humanity "is not the center of the universe physically, but ... may be the spiritual center."[53] Darwin helped heal the "wound inflicted by Copernicus and Galileo." A developmental view of salvation and history could thus be merged between Christianity and science.[54] "If there is no evolution, then all is futility," he wrote in *Genetic Diversity and Human Equality*, but "if the world evolves, then hope is possible."[55] Evolution provides hope that, "while the universe is surely not geocentric, it may conceivably be anthropocentric."[56] A fluid world is a redeemable world, one that may be on the way to deification.[57] Humanity, after all, is "not a passive witness but a participant in the evolutionary process."[58]

But Dobzhansky needed more. He desired a synthesis, and this explains his turn to Pierre Teilhard de Chardin, for whose work Dobzhansky evinced genuine enthusiasm, even as most scientists followed Peter Medawar's scathing review and dismissed Teilhard's *The Phenomenon of Man* out of hand (Medawar termed it "anti-scientific," "unintelligible," and reading it occasioned "real distress, even ... despair").[59] Nevertheless, Dobzhansky was a devoted proponent, to the point that he became president of the Teilhard Association in 1969.

51   Van der Meer, "Theodosius Dobzhansky," 105–112.
52   Ruse, "The Problem of Progress," 240.
53   Dobzhansky, *Mankind Evolving*, 346.
54   Dobzhansky, *Mankind Evolving*, 346.
55   Theodosius Dobzhansky, *Genetic Diversity and Human Equality* (New York: Basic Books, 1973), 113.
56   Dobzhansky, *Mankind Evolving*, 7.
57   Van der Meer, "Theodosius Dobzhansky," 108.
58   Dobzhansky, *Ultimate Concern*, 137.
59   Peter Medawar, "Critical Notice," *Mind* 70:277 (January 1961): 99–106.

Teilhard's thinking offered Dobzhansky the framework of a synthesis. In *Mankind Evolving*, Dobzhansky wrote that humanity needed a faith, a hope—"nothing less than a religious synthesis ... grounded in one of the world's great religions, or in all of them together."[60] He was attracted to Teilhard's developmental and progressive view of history, praising him as "the evolutionist who had the courage to predict future transcendences, mankind moving toward what he called the megasynthesis and toward Point Omega, this last being a symbol for God."[61] Dobzhansky maintained that Christianity was "basically evolutionistic," and necessitated a progressive, linear history rather than a cyclical one ("Creation, through Redemption, to the City of God").[62] Augustine, he argued, "expressed this evolutionistic philosophy most clearly."[63] Cyclical views of history were, in Dostoevsky's words, a "devil's vaudeville," but Christianity's affirmation of time and history meant it could harmonise with evolution.[64] Both Christianity and evolution showed that creation "is an ongoing process, not an event of a distant past." Teilhard pointed to a possible way this synthetic evolution might happen, and Dobzhansky tried to rescue him on orthogenesis, arguing that Teilhard did not really believe in that form of evolution, as his critics maintained.[65]

Naturally, traditionalist critics have not taken too kindly to Dobzhansky's views. Seraphim Rose, in his posthumous *Genesis, Creation, and Early Man*, attacked Dobzhansky not only for his beliefs, but also his practice. He condemned him for not often going to church, and for cremating his wife's body and scattering her ashes in the Sierras. Rose noted with alarm that Dobzhansky gave the commencement address at St Vladimir's Seminary in 1972, and that the seminary had conferred upon him an honorary doctorate. Rose stated Dobzhansky's beliefs were "the usually liberal Christian ideas that Genesis is symbolical" and that

---

60   Dobzhansky, *Mankind Evolving*, 109.
61   Dobzhansky, *Genetic Diversity and Human Equality*, 109.
62   Dobzhansky, *Mankind Evolving*, 112-113; *Ultimate Concern*, 112.
63   Dobzhansky, *Mankind Evolving*, 2.
64   Dobzhansky, *Genetic Diversity and Human Equality*, 111.
65   Dobzhansky, *Mankind Evolving*, 347.

humanity could "cooperate with the enterprise of creation."⁶⁶ Dobzhansky never corresponded with Rose, but he likely would have replied, as he stated in *The Biology of Ultimate Concern*, that the "Fathers of the Church did not always hold views which would at present be described as fundamentalist."⁶⁷ And perhaps he would have argued that Rose's scientific views were as modern as his were, as Rose's were derived almost entirely from the work of Henry Morris and the Protestant fundamentalist world of the Institute for Creation Research. Likely, though, he would not have given Rose much thought. Dobzhansky once tried to change the views of creationist Frank Lewis Marsh, only to eventually throw up his hands and admit defeat at the prospect of changing minds. Though Dobzhansky admitted some respect for Marsh's knowledge of contemporary science, he nevertheless described it later as a "futile and exasperating correspondence." "Discussions and debates with such persons," he wrote, "are a waste of time."⁶⁸

Despite his frustrations with creationists, however, Dobzhansky adopted the label himself, perhaps in an attempt redeem it and wrest it away from antievolutionists. "I am a creationist *and* an evolutionist," he wrote (emphasis original). This is not a label most scientists would dare self-apply, but it is arguably his most synthetic statement, as he wrote in his most famous essay—"Nothing in Biology Makes Sense Except in the Light of Evolution"—a classic whose title mirrors its thesis.⁶⁹

Throughout his life and work, Dobzhansky was the great synthesiser, one who sought to merge the various strands of his interests and life to combine natural selection and genetics, democracy with genetics and Christian ethics, and religion with science. He had saved Darwin, but he worried deeply about the questions Dostoevsky raised

---

66   Seraphim (Eugene) Rose, *Genesis, Creation, and Early Man: The Orthodox Christian Vision* (Platina, CA: St Herman of Alaska Brotherhood, 2011), 573–577.
67   Dobzhansky, *Ultimate Concern*, 112. He attributed this to his reading of Robert T. Francoeur's work *Perspectives in Evolution* (Baltimore, MD: Helicon, 1965).
68   Dobzhansky, *Ultimate Concern*, 96. For a history of this exchange, see Ronald L. Numbers, *The Creationists, Expanded Edition: From Scientific Creationism to Intelligent Design* (Cambridge, MA: Harvard University Press, 2006), 151–153.
69   Theodosius Dobzhansky, "Nothing in Biology Makes Sense Except in the Light of Evolution," *The American Biology Teacher* 35:2 (1973): 125–129.

regarding ethics and morality, and therefore sought to merge his scientific views with his religious ones in order to protect human dignity and equality. Whether or not he was successful is beyond the scope of this analysis, but his work is nevertheless to be admired for its earnestness and ambition. And it must be recalled that the man whom Stephen Jay Gould called "the greatest evolutionist of our century" was an Orthodox Christian, albeit of a rather peculiar style.[70]

In all these realms, it was synthesis that was Dobzhansky's greatest legacy. He worried in *Mankind Evolving*, along with Albert Schweitzer, that "our age has discovered how to divorce knowledge from thought," and he hoped to find ways to mend the breach, stating "attempts to synthesize knowledge are indispensable."[71] Fighting the balkanisation of education, the splitting of philosophy and science, and the hermetic sealing of spirituality from biology, Dobzhansky hoped to find the middle way. The clearest summation of these attempts might have come a mere two years before his death, when Dobzhansky wished to remind everyone that "Evolution is God's, or Nature's, method of Creation. Creation is not an event that happened in 4004 BC; it is a process that began some 10 billion years ago and is still under way."[72]

The author reports there are no competing interests to declare.
Received: 21/03/22 Accepted: 23/08/22 Published: 30/08/22

---

70 Stephen Jay Gould, "Darwinism Defined: The Difference Between Fact and Theory," *Discover* 8, no. 1 (1987): 65.
71 Dobzhansky, *Mankind Evolving*, xi
72 Dobzhansky, "Nothing in Biology Makes Sense Except in the Light of Evolution," 127.

From Physics to Metaphysics: A New Way

Stephen Ames

Abstract: Brian Cox, at the end of his fifth episode in the 2021 BBC series *Universe*, says that big questions like, "why is there anything at all?" are scientific questions about nature. The paper challenges this form of naturalism by drawing on the work of V. J. Stenger, who derived virtually all the great laws of physics L from some physical knowledge and from a principle of point-of-view-invariance used by physicists in their enquiries. We will call this result R. The move from R to metaphysics is motivated by R having the oddity that L, operating from the Big Bang, are derivable from premises that include something that appears billions of years later, namely physicists using the above principle. The move is only justified if it can overcome two blockers: #1 that R is explicable wholly within the resources of the natural sciences; #2 that R is a brute fact. Either way, seeking a further explanation is not justified. I show these blockers logically cannot hold. Seeking a metaphysical explanation of R is therefore justified. It is shown that it is not unreasonable to conclude the universe is structured according to the laws of physics by God, the creator of the universe *ex nihilo*, in order that the universe be knowable through empirical enquiry, by embodied rational agents, using the principle of point-of-view-invariance.

Keywords: laws of physics; physicalism; point-of-view-invariance; *Universe* (2021 BBC series)

Stephen Ames is an Honorary Fellow of the School of Historical and Philosophical Studies in The University of Melbourne, Australia. He holds a PhD in Physics and a PhD in Philosophy of Science, both from Melbourne. The author has greatly benefited from conversations with Neil Thomason, Keith Hutchison, Kristian Camilleri, William Stoeger SJ, Roger Lewis, John Pilbrow, Matthew Pinson SJ, and Sean Devine. Any remaining errors are entirely the author's responsibility.

Throughout my lecturing career, I have encountered several matters that make it difficult for many students to even grasp a Christian account of the scientific view of the universe. One is the sense that the Christian Bible is out of date for anyone with a scientific view of the world. Another is the problem of natural evil, that is, all the pain and death brought about by natural processes such as tsunamis, genetic disorders, the evolution of life on the planet, where such processes are supposedly created by a loving God. Another is the pervasive naturalism of modern culture. Naturalism is the doctrine that nature is all there is. Scientific naturalism says that nature answers to all the objects, relationships, and processes that are identified in the well-established natural sciences.[1] Finally, students would like, if not a proof of God, then, a sense that there are rational grounds for belief in God, especially given pervasive naturalism and the exciting and relentless expansion of the natural sciences, especially physics and cosmology. Our culture is saturated by the natural sciences, technology, and the free market economy. Many people absorb from this milieu the view that there is no purpose or moral order written into the universe, and nothing beyond the universe. Here I draw on what Charles Taylor calls the "immanent frame,"[2] meaning that many people envisage living a good life without any reference to anything transcendent, and get on living it.

In this paper I address two of these issues; pervasive naturalism, and the sought-after rational grounds for belief in God. Naturalism doesn't necessarily present itself in philosophical terms.[3] An example

---

[1]   E. B. Davis and R. Collins, "Scientific Naturalism," in G. B. Ferngren, *Science and Religion: A Historical Introduction* (Baltimore: John Hopkins University Press, 2002), 322.

[2]   C. Taylor, *A Secular Age* (Cambridge, MA: Belknap Press of Harvard University Press, 2007), 589. See also ibid., 548, 566.

[3]   The most philosophically developed form of scientific naturalism is physicalism. David Papineau, "The Rise of Physicalism," in *The Proper Ambition of Science*, ed. M. W. Stone and J. Wolfe (Routledge: London, 2000); David Stoljar, 'Physicalism', *Stanford Encyclopaedia of Philosophy* at http://plato.stanford.edu/entries/physicalism/ (2001); James Ladyman and Don Ross, *Everything Must Go: Metaphysics Naturalised* (Oxford University Press, 2007). As well as defenders of physicalism, there are its critics. C. Hemple, "Reduction: Ontological and Linguistic Facets," in *Essays in Honour of Ernst Nagel*, ed. S. Morgenbesser et al. (New York: St Martin's Press, 1970). See Papineau, "The Rise of Physicalism," 183

is the conclusion by Brian Cox in the last episode of his excellent BBC series, *Universe*. The first episode explores our cosmic origins examining how stars bring meaning to the universe. The second explores whether we are alone in the universe. The third tells how a new space mission has uncovered the history of the Milky Way. The fourth is about the super massive black hole at the centre of our galaxy. The fifth asks why we are here. This episode journeys back 13.8 billion years to the origin of the universe.

At the end of the fifth episode, Cox tells us four things. First, at some length he tells us that scientific enquiry is amazing, given the breadth, depth, and detail of its discoveries about our universe. As a crucial example, he highlights the cosmic microwave background radiation—the most ancient light in the universe. He also notes how much we have learned, though we are located on the tiny speck of our planet in this vast universe. Second, he identifies big questions like "why does anything exist?" and "why do we exist?" Cox grants that to many people these don't sound like questions for science. They are more like questions for philosophy and perhaps even theology. But, third, Cox thinks they *are* scientific questions because they are questions about nature, which we can only answer by looking outwards, beyond the stars, not by looking within ourselves. Fourth, as we engage the universe, we not only ask questions, but we also begin to find answers, by which he means scientific answers.

Cox's assurance that science can provide an answer to the big questions such as "why does anything exist?" is surprising. A couple of years ago, my atheist colleague Dr Kristian Camilleri and I were saying to a class in "God and the Natural Sciences" that if your question is "why is there anything at all?" science won't help you with an an-

---

for his response to Hemple. See also J. Haught, *Is Nature Enough? Meaning and Truth in the Age of Science* (Cambridge University Press, 2006); C. Cunningham, *Darwin's Pious Idea: Why the Ultra-Darwinists and Creationists both Get It Wrong* (Grand Rapids, MI: Eerdmans, 2010); S. Ames, "The Rise and Consequences of Scientific Naturalism," in *Anthropos in the Antipodes*, ed. R. Horner, P. McArdle, and D. Kirchhoffer (Melbourne: Mosaic Books, 2013); S. Ames, "Critique of Daniel Dennett's, *From Bacteria to Bach and Back: The Evolution of Minds*," *Journal of Bioscience & Bio Engineering* 3:1 (2022): 1–7.

swer. Straightaway a young man shot up his hand and said, "you mean science hasn't yet provided an answer." This second-year student was deeply into mathematics and physics. We affirmed the distinction he was making, but not its application in this case. Our claim was not based on a gap in scientific understanding, to be closed by further research. Our claim was based on the fact that any scientific answer necessarily draws on what already exists to do the explaining. Logically, it is unable to explain why there is anything at all. The student accepted this answer and even laughed. It is not a deep or complex point. Of course, we acknowledged that in making this point we were neither claiming nor denying that there is an answer to the question. Everyone knew that Kristian and I have different answers to that question. We left the question open for students to consider. Our point was simple, and it struck me that this student had reached second year university without this having been pointed out before. Doubtless he was not alone.

In what follows I accept Cox's views about where to *start* to seek answers to the big questions, namely the amazing breadth and success of scientific enquiry. This will lead to a critique of the pervasive naturalism of contemporary culture, but not by rehearsing the familiar discussions about physicalism, which shows the need of an ontology richer than that assumed by scientific naturalism. Instead, a new way to make the journey from physics to a richer metaphysics is presented, using the work of physicist and atheist Victor Stenger. In daily talk, people do not make recourse to metaphysics, they rather tell stories. But every story told (or play performed, or movie made) is set within some world and will carry indications of the *kind of world* it is in which the story unfolds. For the story, this is reality. Here, metaphysics is a worldview. It is an account of reality and perhaps some idea of how we know it.[4]

---

4   For a technical account of the meaning of metaphysics, see Neil Omerod, "Bernard Lonergan and the Recovery of a Metaphysical Frame," *Theological Studies* 74 (2013): 960–982, https://doi.org/10.1177/0040563913074004112. Ormerod (ibid., 963) returns to Aristotle's distinction between metaphysics as first philosophy and other "sciences" such as mathematics and physics. Cf. Aristotle, *Metaphysics* 4.1, 10003a24. See also J. Loux, *Metaphysics: A Contemporary Introduction* (Oxford: Clarendon, 1998).

In summary, my approach starts from the relentless expansion of the natural sciences and voices a disciplined speculation based on this very successful form of human enquiry. I will show that the speculation entails two unavoidable questions: "why is there anything at all?" and "why is what there is structured—and structured the way it is?" The evidence for this speculation comes from finding answers to these two questions, which support each other and survive strong challenges.

A Speculation

The speculation is based on three observations about human enquiry. First, any particular research in the natural sciences presupposes that what is being enquired into is intelligible and open to rational explanation, though without prejudice to the forms of intelligibility and the forms of rationality that may be called for. This presupposition is what gets enquiry going and keeps it going. Second, history shows the incessant character of human enquiry, especially the last 450 years of scientific research that continues providing explanations of more and more of the universe in completely natural terms. Third, human enquiry conducts itself and envisages itself as continuing. It does not envisage itself as coming to an end. Human enquiry begins from wonder and proceeds through the continuing eruption of questions on a quest for a true understanding of whatever it researches. The natural sciences powerfully exemplify this dynamic process. Even if institutions (secular or religious) suppress enquiry, questions continue to erupt!

Let us recognise these aspects of human enquiry by the speculation that "all there is, is fully intelligible." Of course, the speculation may lead nowhere—it might prove to be nonsense, or lack any interesting consequences, or there may be no evidence for it beyond the above motivation, and much against it.

Some clarifications are called for and some challenges are noted. Our speculation does not entail that everything is fully intelligible to us now. Human enquiry will never be faced with a brute fact for which there is no explanation. Furthermore, enquiry is not faced with an infinite regress of explanations of the way things are, for then the fully intelligible becomes unintelligible. There are at least three ways the proposition can be challenged. First, a direct challenge is the open ontological question, "Is all there is fully intelligible? After all, the universe may be a brute fact." But do we not risk falling into a gaps argument if we assert that something is a brute fact, when without a larger argument all we can mean is that we have not yet filled the gap in our explanation?

While this proposition does not entail that everything is fully intelligible to us now, it does lead us to expect there ought to be answers for at least the two big questions mentioned above: "why is there anything at all?" and "why is the universe structured—and structured the way it is?" The speculation that all there is is fully intelligible cannot be fulfilled if there is only an infinite regress of explanations. It can only be fulfilled if there is something that explains the existence of everything else, the very nature of which explains its existence, which is to say its existence does not depend on anything else, but rather it exists necessarily. This is the idea of God, the creator of all there is *ex nihilo*—that is to say, not from preexisting stuff.[5] Such a God would have complete understanding, including self-understanding and being self-explanatory. As Ward comments, "being self-explanatory, after all, does not entail that anyone else can understand the explanation, only

---

5   With some differences, here I am very much influenced by B. Lonergan, *Insight: A Study of Human Understanding*, ed. F. E. Crowe and R. M. Doran (Toronto: Lonergan Research Institute of Regis College and University of Toronto Press, 2000), chs 19–20; B. Lonergan, "The General Character of the Natural Theology of *Insight*," in *Philosophical and Theological Papers 1965–1980: Collected Works of Bernard Lonergan*, vol. 17, ed. R. C. Croken and R. M. Doran (Toronto: Lonergan Research Institute of Regis College and University of Toronto Press, 2004), 1–10; B. Lonergan, *Method in Theology* (New York: Herder and Herder, 1972), 101–103; R. Spitzer SJ, *The Soul's Upward Yearning: Clues to Our Transcendent Nature from Experience and Reason* (San Francisco: Ignatius Press, 2015), ch. 3 and Appendix 2; K. Ward, *Rational Theology and the Creativity of God* (Oxford: Basil Blackwell, 1982); K. Ward, "God as a Principle of Cosmological Explanation," in *Quantum Cosmology and The Laws of Nature*, ed. R. J. Russell, N. Murphy, and C. J. Isham (Vatican City State and Berkeley, CA: Vatican Observatory Publications and the Centre for Theology and the Natural Sciences, 1996), 247–262; K. Ward, "God as the Ultimate Information Principle," in *Information and the Nature of Reality: From Physics to Metaphysics*, ed. P. Davies and N. H. Gregersen (Cambridge University Press, 2010),282–300, https://doi.org/10.1017/CBO9781107589056.

that it exists."⁶ Nor, I would add, does it entail that no one can ever come to understand the explanation. Lawrence Krauss concedes that if God is understood as the cause of all causes, then there is no regress of explanations.⁷ Our argument understands God as the cause of all causes and will go on to address Krauss' further claim that there is no evidence for the idea of God.

Here is the beginning of an answer to the first question: "why is there anything at all?" It is a beginning of an answer given that, for example, the claim that God exists necessarily has been criticised on the grounds that a God existing necessarily cannot but act necessarily, including creating necessarily. This necessity excludes freedom from the act of creation and from what is created. This would contradict the freedom manifest in human living, including human enquiry. It would also contradict any idea of God creating freely. This well-known difficulty is noted by Ward⁸ and Paul Davies.⁹ The latter sees this as a fatal difficulty for the idea of God, citing Ward, but without considering Ward's extensive answer to this difficulty in the last chapter of his *Rational Theology*.

Help with this difficulty is also given by Peter Laughlin,¹⁰ who discusses divine necessity and created contingence in Aquinas. A key point for Laughlin is what kind of necessity is meant when God is said to be necessary. For example, did Aquinas intend "logical necessity" when he spoke of God being necessary? Laughlin shows that this is not the case. The problem we are discussing comes from assuming "that if God is the first and necessary cause then there can be no contingent

---

6   Ward, *Rational Theology*, 8.
7   L. M. Krauss, *A Universe from Nothing: Why There is Something Rather Than Nothing* (New York: Free Press, 2012), 167, 170. Here, Krauss concedes that if God is understood as the cause of all causes, then there is no regress of explanations.
8   Ward, *Rational Theology*, 7-8.
9   P. Davies, *The Goldilocks Universe: Why is the Universe Just Right for Life?* (London: Allan Lane, 2006), 231; P. Davies, "Universe from Bit," in *Information and the Nature of Reality*, 66.
10   P. Laughlin, "Divine Necessity and Created Contingence in Aquinas," *The Heythrop Journal* (2009): 648-657, https://doi.org/10.1111/j.1468-2265.2009.00476.x. Laughlin's article is also highly influenced by Lonergan's work Grace and Freedom as a reading of Aquinas on these issues.

proximate causes and *ipso facto* there are no contingencies." The assumption is that whatever comes from, or is brought about by a necessary being, proceeds necessarily (so Neoplatonism). Laughlin argues this assumption is not a problem for Aquinas, for whom creation "is not logically necessary since the proposition 'God does not create' does not by itself entail a contradiction. Indeed, creation is not required by some ineluctable logic or by the nature of deity so that God could not have willed not to create." Rather, if it is open to God to choose between creating and not creating, once having created, it is no longer open to God not to create. "Whatever God wills, then, in the act of willing cannot be changed but God's will remains free to choose what it is that God will in fact will. The acts of God's will are thereby only conditionally necessary in this sense, they are not absolutely necessary for God."[11] Laughlin concludes by quoting Aquinas's point that no absolute necessity can be inferred from the divine will.[12]

Based on our speculation, an answer is also to be expected to the second question, "Why is the universe structured—and structured the way it is?" A reasoned answer is possible only when some idea of how the universe is structured is identified. Many will think of the laws of physics as at least part of the answer and so, in part, our question becomes, "Why is the universe structured according to the laws of physics?" An answer may be reached starting from the work of Victor J. Stenger.

Physics according to Stenger

V. J. Stenger, especially his 2006 book, *The Comprehensible Cosmos*,[13] derives the laws of physics for classical physics, relativistic physics (special and general), quantum mechanics, the standard theory of particle physics, and statistical mechanics.[14] The laws are well known. What is of interest for us here is in how he pursues the derivations.

11   Laughlin, "Divine Necessity," 654.
12   Laughlin, "Divine Necessity," 655.
13   V. J. Stenger, *The Comprehensible Universe: Where Do The Laws of Physics Come From?* (New York: Prometheus Books, 2006).
14   See the table of the basic laws of physics in Stenger, *The Comprehensible Universe*, 113-114.

Stenger starts by considering the kind of objectivity physicists seek in making models of reality. He illustrates this by contrasting the observations physicists make to observations from a subjective point of view, such as taking a photograph. "Instead, physicists seek *universality*, formulating their laws so that they apply widely and do not depend on the point of view of any particular observer. In that way, they can at least hope to approach an accurate representation of the objective reality that they assume lies beyond the perceptions of any single individual."[15] This claim is supported by a brief sketch of science's history of increasing objectivity from Galileo to Einstein. Here, objectivity means that what is observed is not dependent on the position or reference frame of the observer. "This does not mean that the Universe looks the same at every point of space and time." Rather, "while all phenomena may not look the same in detail, they can be modelled in terms of the same underlying principles."[16] Stenger's key idea is this: "Physics is formulated in such a way to assure, as best as possible, that it does not depend on any particular point of view or *reference frame*. This helps make possible, but does not guarantee, that physical models faithfully describe an objective reality, whatever that may be." He claims that when our models are the same for all points of view, "then the most important laws of physics, as we know them, appear *naturally*." A model "should be able to successfully describe in a repeatable, testable fashion a whole class of observations of the same general type; enable the predictions of other unexpected observations; and provide a framework for further applications, such as in technology or medicine."[17]

The key idea amounts to the principle of point-of-view invariance (hereafter, *PPOVI*): "*Point-of-view invariance:* The models of physics cannot depend on any particular point of view."[18] Stenger readily shows that this principle requires the description of reality as invariant to the translation of the origin of the spatial coordinate system (space-translation), the rotation of a spatial coordinate-system (space-rotation),

---

15 Stenger, *The Comprehensible Universe*, 15, 55, 65.
16 Stenger, *The Comprehensible Universe*, 56, 157–159.
17 Stenger, *The Comprehensible Universe*, 9, 10, 15.
18 Stenger, *The Comprehensible Universe*, 57.

and the translation of the origin of the time variable (time-translation). He also designates invariance as symmetry, for example a sphere is invariant under rotation about any axis.[19] Stenger shows that conservation of energy follows from time-translation invariance, conservation of linear momentum follows from space-translation invariance, and angular momentum is conserved by any space-rotation invariance. The conservation laws "are simple consequences of the symmetries of space and time," or, equivalently, "from point-of-view-invariance" using space and time as a framework for constructing models that have invariance under time-translation, space-translation, and space-rotation. Stenger asks:

> where does point-of-view invariance come from? It comes simply from the apparent existence of an objective reality—independent of its detailed structure. Indeed, the success of point-of-view invariance can be said to provide evidence for the existence of an objective reality . . . If we did not have an underlying objective reality, then we would not expect to be able to describe observations in a way that is independent of a reference frame.[20]

If symmetry is the star performer of twentieth century physics, "broken symmetries" are no less important. Stenger discusses symmetry violations, arguing broken symmetry is a fundamental fact about the universe.[21] He counts broken symmetries as a good thing, "at least from a human perspective. Without this complexity and diversity, the Universe would be a dull place indeed, and furthermore we would not be here to be bored by it."[22]

From *PPOVI* and other assumptions and principles (e.g., Noether's Theorem[23]), Stenger elegantly derives all the laws of classical, relativistic and quantum physics (Mathematical supplements A to G).

---

19   Stenger, *The Comprehensible Universe*, 57.
20   Stenger, *The Comprehensible Universe*, 187. In my opinion, this is a hint of metaphysical realism underlying *PPOVI*.
21   Stenger, *The Comprehensible Universe*, 97–106.
22   Stenger, *The Comprehensible Universe*, 102.
23   Stenger, *The Comprehensible Universe*, 58.

This is an impressive *tour de force*. Stenger is clear: "The principle of point-of-view-invariance . . . is an eminently testable, falsifiable principle. So far, it has not been falsified."[24] Nothing guarantees the agreement. The universe might have turned out to be otherwise.

Significantly, Stenger does not claim to derive *all* the laws of physics, such as the second law of thermodynamics, which says that the entropy of an isolated system must remain constant or increase with time. He points out that a broken vase does not reassemble itself. It is not a universal law of physics.[25] It holds at the macroscopic level, describing the average behaviour of systems of many particles, but not at the molecular level and below (atomic, nuclear, subnuclear).

This *PPOVI* concerns the models of reality physicists produce and are consistent with the kind of objectivity they seek. These models cannot depend on any particular point of view. The models are then to be tested empirically. This is a principle about model construction and testing. It is an epistemic principle, guiding physicists' enquiries into the universe. Physicists and their construction and testing of models are an essential presupposition of this principle. The principle does not specify any model, but rather governs the production of any model. Thus, this principle is not reducible to some actual model of reality that meets the requirement stated by the principle, for example a model possessing certain kinds of symmetry.

I accept Stenger's derivation of the laws of physics shown in his supplements A to G, and now want to draw conclusions from this part of his work. The derivations (not just the conclusions) may be gathered and represented as *R: PPOVI, AOA => L*. AOA stands for "all other assumptions" (e.g., about time, space, and matter), which Stenger uses in his arguments to derive the laws of fundamental physics *L*. The *L* are the conclusion to Stenger's argument, but *R* is needed to represent the whole argument. After all, these derivations are what are distinctive about Stenger's work. The derivations show that the fundamental laws

---

24 Stenger, *The Comprehensible Universe*, 161.
25 Stenger, *The Comprehensible Universe*, 21–22, 117.

of physics appear to conform to *PPOVI*. As noted, nothing guarantees the agreement. The universe might have turned out to be otherwise.

The subtitle of Stenger's book asks, *Where do the laws of physics come from?* The derivations already discussed do not answer this question, for they do not explain how the universe appears to have been operating according to these laws from the earliest moments after the Big Bang. To seek help on this subtitle, we turn to his account of the origin of the universe. Stenger's account of the universe's origin sums up physics with the view that the known symmetries are the low energy consequences of the breaking of high energy symmetries. The breaking of symmetries "could be dynamical, that is, the result of some 'lawful' higher process lying still undiscovered." More simply, symmetries could be broken spontaneously, "by a phase transition analogous to the breaking of symmetry when a magnet cools below the Curie point."[26] Symmetry breaking is a violation of *PPOVI*. It corresponds to a particular viewpoint being singled out. In the spontaneous symmetry breaking, the underlying model remains symmetric. Symmetry breaking does not contradict the idea of *PPOVI*.

Exactly what that higher symmetry is still has to be discovered. *PPOVI* simply requires symmetry without specifying any particular symmetry group. Stenger's view is that empirical and theoretical indicators show that supersymmetry (invariance under transformations between bosons and fermions) will likely be part of any future unified model.

Stenger rejects the suggestion that the fine tuning of physical constants for life is the result of an external natural causal agent or "some agency beyond nature" designating a particular set of constants.[27] Nor does he follow physicists who believe that the parameters currently determined by experiment will eventually be derived from some set of basic principles. "It seems highly improbable, however, that any purely natural set of principles would be so intimately connected to the biological structures that happened to evolve on our particular planet." In his view it is more likely that life evolved in response

---

26 Stenger, *The Comprehensible Universe*, 166.
27 Stenger, *The Comprehensible Universe*, 168.

to the physical parameters characterising our universe. Spontaneous symmetry breaking would mean the values of the constants arose by accident. "If we had an ensemble of universes, then the parameter values in our Universe arose from a random distribution—with no external, causal agent designating one particular set." Stenger's view is that the "observable universe, in fact, looks just as it would be expected to look in the absence of any such agent. The laws of physics are . . . 'lawless laws' that do not arise from any plan but from the very lack of a plan. They are the laws of the void."[28]

By *void*, Stenger means a vacuum that has zero vacuum energy. Various possible ways of thinking about zero energy are considered, viz., super-symmetric vacuum: negative energy solutions for the energy field. The issue is "how to get matter from a symmetric void."[29] Stenger appears to offer two answers, which I will not discuss here, in terms of quantum tunnelling and of the collapse of the symmetric void.[30] While I have questions about these answers, I will show that my larger argument has no need to resolve these and other possibilities, including a multiverse. I can happily wait upon these matters to be resolved scientifically.

Moving from Physics to Metaphysics: Can the Move Be Justified?

The Motivation

The theme of this paper is the move from physics to metaphysics and so the motivation for this move is sought from within physics. Previously, the motivation for espousing scientific naturalism was the expanding success of scientific explanations, the basis for a positive induction that every question about our universe will be similarly answered. Here it

---

28 Stenger, *The Comprehensible Universe*, 169.
29 Stenger, *The Comprehensible Universe*, 148.
30 Stenger, *The Comprehensible Universe*, 150, 170.

is found in Stenger's derivations of the form of the laws of physics L, which may be summarised as R: $PPOVI, AOA \Rightarrow L$.

There is an apparent oddity in R. The L, operating since very soon after the Big Bang, is explained in terms of $PPOVI$ which refers to a principle used by enquirers that only show up billions of years later. This seems odd and leads to the question: is R true of the L and so true of the L operating from the earliest moments after the Big Bang? $PPOVI$ yields laws that hold for all viewpoints and reference frames, including those located soon after the Big Bang. If we answer affirmatively, then we may wonder how does it come about that the L operating from the earliest moments after the Big Bang are derivable from premises that nontrivially include $PPOVI$, which refers to physicists conducting their enquiries billions of years later?

From a different angle, anyone working from a strongly naturalistic standpoint may be skeptical about this question, not giving it much weight and certainly not allowing anything to be built on a mere question. This skepticism would aim to show how R can be explained wholly within the resources of the natural sciences and physics in particular.[31] After all, R has been obtained using these resources. If the oddity of R is only apparent, explicable after all in terms of the resources of the natural sciences, there would then be no justification for seeking a metaphysical explanation of R. Call this, blocker #1. Also, if it were reasonable to interpret R as a brute fact and therefore without further explanation, there would be no justification for seeking a metaphysical explanation of R. Call this, blocker #2. It can be shown that the resources of the natural sciences are logically unable to explain R. Blocker #1 is defeated. It can be shown that, logically, it is unreasonable to treat R as a brute fact. Blocker #2 also is defeated.

---

31   E. Carlson and E. J. Olsson, "Is Our Existence in Need of Further Explanation?" *Enquiry* 41:3 (1998): 255–275.

How Blockers #1 and #2 Are Defeated

Blocker #1 seeks a physical theory T_{phys} that explains R. In brief, a physical theory T_{phys} is:

- a "blind" causal explanation of physical events and processes; "blind" means no final causes, goals, purposes built in;
- the causal explanation is described mathematically and aims to derive a mathematical description of what is to be explained;
- open to empirical testing.

Blocker #1 would be T_{phys} => R. A series of problems are foreseeable:

- R is the wrong kind of explanandum for any T_{phys}
- R is a rational inference. It stands in the logical space of reasons, not in the very different logical space of subsumption under natural laws.[32]
- Logically, R can never be obtained from any T_{phys} (as defined).
- T_{phys} has to provide *PPOVI* for the derivation of R to succeed.
- If T_{phys} includes *PPOVI*, then T_{phys} is not "blind." *PPOVI* is about physicists pursuing valued epistemic ends guided by *PPOVI* in some universe, which T_{phys} at least in this way envisages.
- Can T_{phys} lead to *PPOVI*?
- No. Physics alone cannot do this; it took the evolving processes of the 13.7-billion-year-old universe (physical, chemical, biological, and cultural) to bring about the existence of enquirers guided by *PPOVI*.

Conclusion: Any physical theory (so construed) logically cannot explain R. Blocker #1 fails.

---

[32] W. Sellars, "Empiricism and the Philosophy of Mind," in *The Foundations of Science and the Concepts of Psychology and Psychoanalysis*, ed. H. Feigl and M. Scriven (University of Minnesota Press, 1956), 253–329; J. McDowell, "Naturalism in the Philosophy of Mind," in *Naturalism in Question*, ed. M. De Caro and D. Macarthur (Harvard University Press, 2004), 91–105.

Blocker #2 claims it is reasonable to treat R as a brute fact about the universe. Consider the following argument concerning R:

- If no scientific or nonscientific explanation of R is possible, R is a brute fact.
- No scientific theory can explain R.
- No nonscientific explanation of R is possible.
- Therefore, R is a brute fact.

The argument is valid. But if we reject the conclusion, as stated in the final dot point, which of the three preceding premises will we reject?

- R established above.
- Says what is meant by a brute fact.
- This is the failure of blocker #1.
- Says that there is nothing outside or beyond what the natural sciences can tell us, that can explain R.

How shall we assess this last point? An initial question is how do we know that no non-scientific theory can explain R? That would be the case only if we assumed scientific naturalism with its methodological, epistemic, and metaphysical theses. The latter says that all there is is what physics says there is, or complex configurations of the same. But with R we are concerned with something that scientific theories logically cannot explain, something beyond the scope of scientific theories.

PPOVI is obtained initially quite independently of knowing the evolutionary cosmology of the 13.7-billion-year-old universe. It is obtained by rational enquirers, with certain aims and some general beliefs about rationality and about how the world operates deciding what standards rationally ought to be met by actions directed to achieve valued epistemic ends. Analogous considerations have their place in practical actions like shooting an arrow from a bow to hit a target. We know about rationality because human beings instantiate rationality, whereby they think and act for various reasons, but this is known independently of how the origins of that instantiation might be explained.

This is one argument for thinking of *PPOVI* as something beyond the theories of natural science, yet *PPOVI* is nontrivially involved in explaining the form of the laws of fundamental physics L, as shown in R. This provides rational grounds for wondering if something beyond the natural sciences might explain R. But the penultimate dot point would lead us to expect any such explanation to be impossible. Hence the last dot point should be set aside as unreasonable. Therefore, the last dot point does not follow, and we reasonably set aside the claim that R is a brute fact. Note that this result is not based on Leibniz' principle of sufficient reason. Blockers #1 and #2 fail. We are therefore justified in seeking further—beyond the resources of the natural sciences and physics in particular—a metaphysical explanation of R, including the oddity in R.

A Metaphysical Explanation of R

Seeking such an explanation is guided by the question, "What must minimally be assumed to hold to explain R?"

Any explanation of R must provide *PPOVI*. Whatever provides *PPOVI* is something that has language, that has access to the logical space of reasons, and thereby logic and mathematics, and it knows about intentionality—*PPOVI* assumes embodied rational agents (humans or aliens) in a universe (whether our universe only or within a multiverse) pursuing valued epistemic ends concerning that universe.

These are very good grounds for saying that *only* something capable of rational thought can provide *PPOVI*. This "something" should be thought of as some kind of rational agent, "*RA*." A rational agent must be assumed because thought alone is not enough to explain the existence of any universe or multiverse however conceived. To explain how R holds for our universe, we must assume that RA envisages a universe at least for which R holds, as in the preceding paragraph. That is, we must minimally think of RA envisaging a universe at least operating according to *L* and for which *AOA* holds, for which *PPOVI* also holds, and that the universe so envisaged eventually produces embodied rational agents capable of pursuing valued epistemic ends guided by *PPOVI*.

We may properly treat this as the end/purpose *RA* envisages for this universe. This *purposive* explanation arises from within the argument rather than being imposed. (This purposive explanation at least invites the question of whether this end may be included in any larger end *RA* possibly envisages for this universe.) For *R* to be true of an existing universe, *RA* must also be understood as somehow bringing about this envisaged, but so far in this argument, not existing universe. Meeting this requirement would allow the developing explanation to be an answer to the question: Why is the universe structured and structured according to the laws of physics?

If the argument from Stenger's work to this point was all we had to go on, a Kantian note would be that the most we could claim would be that *RA* is the architect of the envisaged universe, to be produced from some pre-existing stuff. We began the argument, however, from a speculation starting from the observation that human enquiry presupposes that what is being enquired into is intelligible and open to rational explanation, but without prejudice to the forms of intelligibility and rationality that may be called for. Based on the relentless expansion of human enquiry that is apparently unending, the speculation generalises that presupposition by assuming that all there is, is fully intelligible. That generalised presupposition blocked the idea of an infinite regress of explanations of the universe and the idea of the universe being a brute fact. The generalised presupposition entailed the expectation of answers to two unavoidable questions: "why is there anything at all?" and "why is the universe structured—and structured the way it is?" Based on Stenger's work, we have the beginning of an answer to the second question. This supports the generalised presupposition and therewith the first question. Earlier we found the beginning of an answer to the first question by arguing to the idea of God, the creator of all there is *ex nihilo*—that is to say, not from preexisting stuff. Should we think that God creates *RA* or identify God as *RA*? If the first, then God must at least already have all the characteristics of *RA*, allowing us to identify God as *RA*. This is the simplest explanation of Stenger's result *R*.

We may conclude that God the creator of the universe *ex nihilo* has structured the universe (at least) in term of the laws of physics in order that the universe be knowable by embodied rational agents (human or alien) though empirical enquiry guided by *PPOVI*.

Discussion

The paper presents a new way of proceeding from physics to metaphysics, largely drawing on a speculation about the universe, based on: the relentlessly expanding success of the natural sciences; the observation that any scientific enquiry presupposes that what is enquired into is intelligible and open to rational explanation; and Stenger's derivation of the laws of physics from premises that include *PPOVI*. Stenger's result has an oddity that the laws of physics operating in the universe including from the earliest moments after the Big Bang are derived from premises that include *PPOVI*, an assumption about what only shows up billions of years later. The oddity could be tested and refuted by showing it can be explained entirely within the resources of physics. It is shown that this testing fails in principle. This critique of scientific naturalism is independent of other criticisms in circulation (see n. 3), and so contributes something new to the literature on scientific naturalism and physicalism in particular.

Generalising the presupposition of human enquiry led to having to face the questions "why is there anything at all?" and "why is the universe structured—and structured the way it is?" Answering the second question began by noting that the laws of physics must surely count as partly identifying how the universe is structured. Drawing on Stenger's work, the argument led to the conclusion that the laws of physics are the way they are in order that the universe be knowable by embodied rational agents conducting empirical enquiries in the light of *PPOVI*. This leads to the expectation of other laws or other ways the universe is structured to bring such embodied agents into existence, and this may be pursued for example together with Daniel Dennett[33] and Paul Davies.[34] This line of thought leads to the expectation of a solution to the hard problem of consciousness, which may be pursued, for example, in conversation with Robert Spitzer[35] and Daniel A. Helminiak,[36] concerning proposed solutions to this problem.

Challenges, Strengths, and Limitations of This Argument

Two important challenges have been raised in discussions. The first claims that my use of *PPOVI* represents a category mistake, because *PPOVI* is a methodological principle guiding research not an ontological principle, making ontological proposals. This claim is correct and concurs with Stenger's thought that if "the models of physicists can be used to successfully describe previous observations and predict future ones, then we can use them without getting into metaphysical questions."[37] It turns out, however, that *PPOVI* can lead to ontological consequences for anyone embracing scientific naturalism. This is shown in my discussion of blockers #1 and #2. The challenge does not attend to this argument justifying the move from physics to metaphysics. In my opinion, there is also a hint of metaphysics in Stenger's view of physicists as seeking "universality," or an "accurate representation of the objective reality that they assume lies beyond the perceptions of any single individual."

A second challenge is that there may be alternative approaches aiming to explain why the laws of physics are the way they are. If so, would Stenger's result be all that significant, when there may be other premises X, such that $X \Rightarrow L$? If this were the case, why build

---

33   D. Dennett, *From Bacteria to Bach and Back: The Evolution of Minds* (Allen Lane, 2017).
34   P. Davies, *The Demon in the Machine: How Hidden Webs of Information Are Solving the Mystery of Life* (Allen Lane, 2019).
35   Spitzer, *The Soul's Upward Yearning*, ch. 6.
36   D. A. Helminiak, *Brains, Consciousness and God: A Lonerganian Integration* (Albany: Suny Press, 2015), chs 4 and 5.
37   Stenger, *The Comprehensible Universe*, 8.

anything based on R? I accept this as a proper concern. The search for contenders for such an X is evident, for example, in the work of P. Davies[38] and Roberto M. Unger and Lee Smolin,[39] though with derivations only as promissory notes. On the other hand, B. Roy Frieden[40] has actually derived many of the laws of physics starting from Fisher information. This is the form of information introduced by R. A. Fisher at Cambridge, in the 1920s, who showed that Darwin's theory of evolution by natural selection and Mendel's genetics made sense statistically. Later, the mathematical form of what came to be called "Fisher information," in honour of Fisher's earlier research, showed up independently in the work of Harald L. Cramer[41] and C. Radhakrishna Rao.[42] They were theorising about how to measure a quantity that is subject to "noise" and so is fluctuating around some mean value θ. It is known as "classical measurement theory." Their celebrated result is the Cramer-Rao Inequality (CRI): $I\,e^2 \geq 1$, where e^2 is "the mean square

---

38 Davies, "Universe from Bit."
39 R. Unger and L. Smolin, *The Singular Universe and The Reality of Time* (Cambridge University Press, 2015).
40 B. R. Frieden, *Science from Fisher Information: A Unification* (Cambridge University Press, 2004); B. R. Frieden and A. G. Gatenby, eds, *Exploratory Data Analysis Using Fisher Information* (London: Springer Verlag, 2007). Frieden's work has been criticised by D. Lavis and R. Streater, "Physics from Fisher Information," *Studies in the History and Philosophy of Modern Physics* 33B:2 (2002): 327–343; for example, that his earlier derivation of quantum mechanics in effect assumed the De Broglie hypothesis. Frieden subsequently showed how the hypothesis can be derived from his "Fisher information" approach to physics. See B. R. Frieden and B. H. Soffer, "De Broglie's Wave Hypothesis from Fisher Information," *Physica A—Statistical Mechanics and Its Applications* 338:7 (2009). A senior physicist, T. Kibble, once required me to provide evidence, independent of Frieden, for thinking there was any fundamental connection between Fisher information and physics. I sent him the following paper which he had not known, but which he conceded that did indeed provide that evidence. S. L. Braunstein and C. M. Caves, "Statistical Distance and the Geometry of Quantum States," *Phys. Rev. Let.* 72:22 (1994): 3439–3443. These brief comments on Frieden's work are drawn from my (unpublished) PhD thesis at the University of Melbourne, 2005, "Cosmology and the Metaphysics of Enquiry: Towards a Non-Materialist Metaphysical Research Programme that Explains and Derives the Fundamental Laws of Nature."
41 H. L. Cramer, *Mathematical Methods of Statistics* (Princeton University Press, 1946).
42 C. R. Rao, "Information and Accuracy Attainable in the Estimation of Statistical Parameters," *Bull. Calcutta Math. Soc.* 37 (1945): 81–91.

error in the measurement-estimates of the fluctuating parameter" and *I* is the "Fisher information."

Of interest is that the approaches of Stenger and Frieden make human enquiry central to the derivation of the laws of physics. Stenger assumes reality exists independently of what human beings know about it and draws the conclusion that physicists' view of the universe cannot be dependent on a particular viewpoint. This is the basis of his *PPOVI*, central to his derivations of *L*. Frieden starts from classical measurement theory to determine the mean value of a fluctuating parameter. This argument is set within the space and time of classical physics. Frieden shows how this leads to "Fisher information" *I*, and the derivation of the Lorentz transformation, with the result that *I* is shown to be invariant and covariant under the Lorentz transformation. This provides a different basis for arriving at point of view invariance. Further comparison of the two approaches would highlight the role of Noether's theorem in Stenger's approach (see n. 17) and "Fisher information" which has the mathematical form of what is called an "action integral."[43] Stenger's result is summarised, *R: PPOVI, AOA => L*, where-

---

43 The mathematical form of Fisher information *I* is called an "action integral." It is natural in the sense that it follows logically from the assumptions from which the Cramer Rao inequality ($I\,e^2 \geq 1$) is derived. These assumptions concern the measurement of a parameter of a system undergoing fluctuations. The measurement proceeds by a probe particle fired at and interacting with the system to be measured. This happens under ideal epistemic conditions (e.g., no noise from the measurement system; see Frieden, *Science from Fisher Information*, 98). In this context and from other properties of Fisher information *I*, Frieden forms another action integral *K* characterising the measurement interaction. Frieden postulates that *K* has the property that an infinitesimal variation of *K*, denoted by δ*K*, is zero, i.e., δ*K* = 0. To put the matter briefly, δ*K* = 0 allows Frieden to use the rich mathematical resources of Lagrangian Mechanics (so named after famous French mathematician Joseph-Louis Lagrange, 1736–1816). The use of these resources leads to second order differential equations of the kind we see in the laws of physics. This is the basis for Frieden's derivations of many of the laws of physics. The extremum principle δ*K* = 0 is also a symmetry principle and so makes connections to Noether's Theorem mentioned earlier. See Frieden, *Science from Fisher Information*, 3 for an important comment on the use of Noether's Theorem. For standard texts on the physics and mathematics, see J. B. Marion and S. T. Thornton, *Classical Dynamics of Particles and Systems* (Fort Worth: Saunders College Publications, 1995), 214–217; H. Goldstein, *Classical Mechanics* (Massachusetts: Addison-Wesley, 1959), 37–38.

as Frieden's result may be summarised $R_F{:}E_F$, $AOA_F \Rightarrow L$, where E_F represents idealised parameter measurement, AOA_F stands for "all other assumptions," and the subscript F indicates Frieden's approach. That comparison will be for another time, as will comparing any other approaches to deriving the laws of physics, especially as they take account of dark matter and dark energy. A third challenge is based on studies examining whether physical constants vary over time.[44] Stenger's argument has basic physical constants invariant over time, which is still the standard view.

A limitation of the argument in its present stage refers to its theology as undeveloped in several ways. Philosophically, the idea of God entered the argument as an answer to the question "why is there anything at all?" Which is a thread in a larger canvas of natural theology for which I would especially commend Spitzer's *The Soul's Upward Yearning*. It is what allowed me to draw on Aquinas via the work of Laughlin's "Divine Necessity." Spitzer's argument would reframe the idea of God used here, just as it reframes the idea of God as the architect of the universe. This still larger idea of God would call us to engage questions such as what kind of world should we expect God to create.[45] Another limitation (and strength) refers to the fact that the argument leaves open an answer to how the universe was structured the way it is. Part of that answer will be given by physicists working on the physics of this question, and I wonder what theology might contribute. For example, my colleagues wanting to understand how God supposedly create all there is *ex nihilo*. Another limitation is that no appeal has been made to the Christian understanding of the vulnerable yet invincible triune God.[46] This is a methodological limitation because this is where I want to begin to engage those who do not share this or any understanding of

---

44   M. R. Wilczynski et al., "Four Direct Measurements of the Fine-Structure Constant 13 Billion Years Ago," *Science Advances* 6:17 (2020), DOI: 10.1126/sciadv.
45   S. Ames, "Why would God use evolution?" in *Darwin and Evolution in Interfaith Perspectives*, ed. J. Arnould (Adelaide: ATF Press, 2009), 105–126.
46   Among many works, see E. M. Conradie, *The Earth in God's Economy: Creation, Salvation and Consummation in Ecological Perspective* (Zurich: Lit Verlag, 2015).

God, who happily live and work within a naturalistic view of the world and its accompanying narrative.

A strength of the argument is that the conclusion is independent of whatever physicists finally conclude about a multiverse. A consequence of the multiverse idea in its various forms (though not its motivation) is a "Darwinian" style objection to any purposive account of why the universe is structured the way it is. That objection does not apply here since my argument does not depend on rejecting the multiverse idea. A purposive answer to why the universe is structured, and structured the way it is, is arrived at from within the argument, rather than being imposed. This purposive answer does not trouble nor is it troubled by Darwinism. It provides a purposive account of natural laws that undergird the operation of the universe including Darwinian evolution. It means the "Watchmaker" is not blind, though the full purpose of God in creation is not thereby revealed. Allow me to illustrate. The room where I am working is filled with "blind" processes that have been set in place for a range of purposes. This is also true of the blind processes in our universe. (We need to be careful about the inference from blind to purposeless.) The designers of my workspace had their immediate purpose and their ultimate purpose. Even if we could infer the former from the blind processes (back engineering), in order to know the latter we would need the designers to disclose or reveal their ultimate purpose. We have not yet considered any argument for the idea of God having any ultimate purpose, nor for God disclosing or revealing such a purpose for the created universe.[47]

Another strength is that the argument allows an answer to why empirical enquiry by embodied rational agents is so important that it is included within (part of) the purpose for which the universe is created by God. The question returns us to the earlier discussion. While God exists necessarily, but not with logical necessity, God freely creates all there is *ex nihilo*. The created world reflects this freedom. Therefore, pure thought alone will not be able to deduce the correct understanding

---

47 For an indication of such an argument see Ames, "Why would God use evolution?" 112, 116–122.

of the God-given, contingent processes of this universe. To approach that understanding, enquirers will have to investigate the particular processes with their senses. The above argument also leads us to think the created world will reflect the rationality of God, but without prejudice on the part of enquirers to the forms of intelligibility and rationality that might be called for in understanding the world; and, I would add, even more so to do with attempts to understand God. Therefore, enquiry into the universe must be sensory, intelligent, and rational. This goes some way towards characterising empirical inquiry. This argument leaves for another time an account of why God would be interested in such empirical inquiry taking place in this created universe.

Conclusion

This overall argument brings to light an account of divine purpose as immanent in the operation of the universe according to blind natural laws. This argument has nothing to do with Intelligent Design, Anthropic principles, Fine Tuning, nor the old argument *from* design. It is not a "gaps" argument, nor does it entail deism, and makes no use of Leibniz' principle of sufficient reason. It is unaffected by whatever turns out to be physicists' conclusion about the multiverse proposal. This is an argument from physics to metaphysics. It is metaphysics because it goes beyond physics to what physics does not enquire into. It is not a physical explanation, but an explanation of the physical in terms of the purpose for which the laws of physics are the way they are.

It is however a metaphysics of enquiry sustaining the principle of point-of-view invariance. Given its key result, it logically cannot conflict with empirical enquiry. This argument is certainly not a science stopper! It logically cannot inhibit either empirical or theoretical enquiry in physics or any other science. On the contrary, it strongly encourages the continuing exploration of both physics and metaphysics as deeply in accord with why the universe is the way it is.

Brian Cox rightly praises the scope and detail of our scientific knowledge of the ... planet. While he acknowledges this contrast, the contrast does not itself lead to any wondering about how this is possible. Presumably, this is because the scope of scientific methods of enquiry and the empirical vindication they offer is well known. The contrast between the speck and its vast context does lead to big questions, such as "why is there anything at all?" and "why are we here?" Cox takes these as questions about nature and as scientific questions, as if there are no other kinds of questions about nature. This paper offers an answer to these big questions, not a scientific answer, but a metaphysical one entirely friendly to the sciences.

Victor Stenger derived a great many of the great laws of physics, and the derivation entailed an oddity. This paper identifies and explains the oddity, after showing that the natural sciences logically could not explain it. Another way of stating the oddity is that the people telling the scientific story of the universe cannot be properly located within the story. Stenger also cited the famous statement of Einstein, that the most incomprehensible thing about the universe is that it is so comprehensible. This paper begins to indicate how we might make the stunning comprehensibility of the universe comprehensible.

The author reports there are no competing interests to declare.
Received: 06/06/22 Accepted: 08/09/22 Published: 15/09/22

Faith, Deuteronomy 18:21–22, and the Scientific Method

Charles Riding

> **Abstract:** This article shows that beliefs or convictions permeate the use of the scientific method just as they permeate religion. To that end, it begins by showing how belief is a prerequisite for both religion and for the deployment of the scientific method as a valid tool for empirical science. Then it describes the scientific method, bringing to the fore the extent to which it entails faith or beliefs. It also shows that Deuteronomy 18 and other biblical passages prove critical thinking to be embedded in the faith both in the use of religion and in the scientific method.
>
> **Keywords:** circular reasoning; conflict narrative; faith; falsification; scientific method

The conflict narrative posits that religion and empirical science are always in conflict—totally incompatible with each other—poles apart. "Religion is founded on faith; but science is founded only on facts!" is the boast of atheists.[1] They also ask how can there be any faith involved in the use of the scientific method, which draws upon tangible evidence, visible facts, hard data, and physical proof. This article attempts to show that faith—or rather belief, conviction—permeates the use of the scientific method just as it permeates all religions. To that end, I shall discuss belief as a prerequisite for both religion and the

---

Rev. Dr Charles Riding is a retired minister of the Presbyterian Church of Australia, having ministered in six parishes over forty years. Before that he taught mathematics and science, particularly physics, at all high school grades. He has published several articles in *Reformed Theological Review*.

1   Jerry Coyne, "Yes, There Is a War Between Science and Religion" (2018) https://theconversation.com/yes-there-is-a-war-between-science-and-religion-108002 (accessed 16 April 2022).

scientific method, to then show that Deuteronomy 18 and other biblical passages prove critical thinking to be imbedded in the faith. Then I shall describe the scientific method, bringing to the fore the extent to which it entails faith or belief.

Faith

Faith or belief has been defined in many ways, from the New Testament's "faith is the substance of things hoped for, the evidence of things not seen" (Heb 11:1), on the one hand, to Mark Twain's "faith is believing what you know ain't so," on the other.[2] Dictionary definitions refer to "Belief: Acceptance as true of any statement, etc." and "Believe: Trust the word of a person; Put trust in the truth of a statement."[3] The definition adopted here is: "Faith is taking a step beyond what the evidence conclusively proves," which is in line with both *The Concise Oxford Dictionary* and the Bible. There is much more that could be said about faith and, indeed, philosophers have said much, much more. It involves knowledge of, acceptance, or mental assent to something, and acting upon the proposition that is believed. The essential point made here is that faith goes beyond proof.

We can supply reasons, facts, and arguments to support our beliefs, for why we accept certain theories, hypotheses, and statements as true, but the former never prove the latter. Faith always goes beyond evidence. It does so in one of two ways—what will be called here the "step of faith" and the "leap of faith." A "leap of faith" (popularised by existentialism) is to go beyond the evidence in the opposite direction to where the evidence appears to be leading. Existentialists believe that this universe is absurd, that there is no purpose or significance in it because it has no creator. We are merely highly evolved pond scum living amongst other highly evolved pond scum and some not so highly evolved pond scum. Even so, existentialists take a leap of faith to

2 Mark Twain, "Faith" (2015) http://www.twainquotes.com/Faith.html (accessed 20 April 2017).
3 "Belief" and "Believe" in *The Concise Oxford Dictionary*, ed. H. W. Fowler and F. G. Fowler (Clarendon: Oxford University Press, 1964).

believe they are significant and hopefully to create purpose for themselves, even though they know this is absurd. As Francis Schaeffer put it, "Kiekegaard came to the conclusion that ... you achieved everything of real importance by a leap of faith. So he separated absolutely the rational and logical from faith. The reasonable and faith bear no relationship to each other."[4] This is what Mark Twain quipped.

On the other hand, a "step of faith" is going beyond the evidence in the direction that the evidence appears to be pointing. Such steps of faith are often made unconsciously because they appear logical and reasonable. The more supporting evidence we have to believe a person or proposition, the smaller the step of faith needed to believe or put our faith in them.

For example, consider reading a crime or "whodunnit" novel. Suppose the author depicts the murder of a rich, married woman. All of the suspects have many motives for wanting her dead. All of the suspects have alibis for the time of her murder. All of the suspects have a web of relationships and intrigue with each other so that any two or more of them could have hatched a conspiracy to murder her and cover each other's tracks. As the story progresses, all the evidence points to the butler. You might conclude: "I believe the butler did it!" That would be a "step of faith" because it was in the direction the evidence appeared to be pointing. Someone else might say: "I know all the evidence points to the butler, but I still believe her husband/widower murdered her." This would be a "leap of faith," since it is in the opposite direction to where the available evidence is pointing. You then have to wait until the end of the novel to find out who the actual culprit/s is/are.

Atheists maintain that all religious faith is a "leap of faith," saying: "Faith means claiming something to be TRUE without any evidence, and despite evidence to the contrary."[5] In the Bible, faith in Jesus Christ is depicted as a "step of faith." Typically, Jesus says to Philip, "Believe Me that I am in the Father and the Father in Me, or else be-

---

4   Francis August Schaeffer, *The God Who Is There: Speaking Historic Christianity into the Twentieth Century* (London: Hodder and Stoughton, 1968), 20–21.
5   Atheist Max, "Is Atheism a Faith?" (2019) atheistmax.wordpress.com/is-atheism-a-faith/ (accessed 17 October 2019).

lieve Me for the sake of the works [i.e., the evidence] themselves" (John 14:10). Accordingly, Christians suspect that it is atheists who are taking the "leap of faith" because all the evidence in the world around us indicates that there must be a Creator behind it all. Atheists retort that they have explained most things scientifically and will one day explain everything without any need for a mastermind, so it is a "step of faith" to believe that there is no Creator. And the argument goes on. We will have to wait until the end of life to find out the actual truth.

Deuteronomy 18:21-22 and the Criteria of Prophecy

Deuteronomy 18 spells out some of the differences between God's people and the surrounding nations. Having told the Israelites not to be like the pagans who seek soothsayers and the like to determine God's will (Deut 18:9-14), Moses then told them that God would raise up a prophet like himself to guide them (Deut 18:15-20). Furthermore, anticipating the appearance of false prophets who would lead Israel astray, Moses gave the people a way to tell true from false prophets: "And if you say in your heart, 'How may we know the word that the Lord has not spoken?'—when a prophet speaks in the name of the Lord, if the word does not come to pass or come true, that is a word that the Lord has not spoken; the prophet has spoken it presumptuously. You need not be afraid of him" (Deut 18:21-22).

The principle behind testing prophets and their prophecies is captured by the end of the passage. It amounts to considering what the prophets predict, and if that does not happen, then their claims to be prophets of the God of the Bible are illegitimate. They are false prophets and can be safely ignored. In the Bible can be found other tests of prophets, described in Deut 13:1-5, 1 Kgs 18:19, Isa 8:19-20, and Jer 23:14 and 28:7-9. The Bible itself has been proven by this method, as all of its prophets, Jesus Christ included, passed the test described in Deut 18:21-22. They also passed all the other tests.

While Deut 18 is often referred to as a test of true and false prophets, it is, strictly speaking, a test of false prophets. It answers the question: "How may we know the word that the Lord has *not* spoken?" (Deut 18:21; my emphasis). Paraphrasing Karl Popper, it is about the falsification, not the verification, of someone's claim to be a prophet of YHWH.[6] In short, predictions that do not happen are only made by false prophets. But what if the predictions do happen? Does that prove the prophets to be true? Not necessarily. They may be false prophets with a lucky guess. The Bible acknowledges that a true prediction may be given by a false prophet: "If a prophet or a dreamer of dreams arises among you and gives you a sign or a wonder, and the sign or wonder that he tells you *comes to pass* [i.e., the prediction does happen], and if he says, 'Let us go after other gods' … you shall not listen to the words of that [false] prophet" (Deut 13:1–3a; my emphasis).

The reason why this test can only falsify a prophetic claim, not verify it, is because the test entails circular reasoning: it starts with the prediction and finishes by comparing what happens with that prediction. Circular reasoning can only prove if a proposition is consistent, not whether it is consistently right or consistently wrong. To determine the rightness or wrongness of a statement, another test or more tests are required, including a step of faith. In the case of the Bible, the next step or the next test of prophets and prophecies requires to ask whether the prophets and the prophecies agree with the teaching of the rest of the Bible. In Isaiah's words: "To the law and to the testimony! If they [the prophets, etc.] do not speak according to this word, it is because there is no light in them" (Isa 8:19–20; see also Deut 13:1–16). The step of faith involved here is believing that the Bible is accurate and reliable, and using it to test potential prophets and prophecies.

A step of faith is required even before using this test. Before applying it, one needs to believe it is a valid test to use. The reasons for accepting it as valid are irrelevant. One may accept it because of believing the Bible is inspired and infallible. Or one may believe it because it sounds logical—or for any other reason/s. But believing it is appropriate is a prerequisite for using it.

---

6 Karl Raimund Popper, *The Logic of Scientific Discoveries* (Mansfield Centre, CT: Martino Publishing, 2014), 32–40.

This test also implies that a prophecy must be falsifiable—that the opposite of the prediction might happen. Some prophecies are so vague or so ambiguous that they will always seem correct. Such predictions are useless, however plausible and religious they sound. Whatever happens, the prophecy makes no difference one way or the other.

Furthermore, prophecies can originate from anywhere. James Crenshaw examined many proposed tests of prophets and prophecies. One such is the "revelatory form" by which the prophet received God's message for the people—whether by dream, by vision, by the word of YHWH, or by the spirit of YHWH, concluding that such "revelatory forms" provide no criterion for distinguishing a true prophet from a false prophet.[7] The most common means of revelation to prophets in the Bible were hearing God's Word (e.g., Jer 1:4, 7; Ezek 3:18; 7:1; Zech 4:8; 8:9) and/or seeing God's message in a vision (e.g., Isa 1:1; 2:1; Ezek 1:3-4; Obad 1). Usually the prophets would then preach it to the people, but occasionally they would act it out (Isa 20:1-6; Ezek 4:1-8). However, God's message sometimes came through other means, such as the "common events" that happened around them. Here is an example: "As [the prophet] Samuel turned around to go away, [King] Saul seized the edge of his robe, and it tore. So Samuel said to him, 'The Lord has torn the kingdom of Israel from you today, and has given it to a neighbour of yours, who is better than you'" (1 Sam 15:27-28).

One of the tests to be used of prophets is that we should expect Godly character from God's prophets, whereas false prophets often live immoral lives (Jer 23:14; 2 Pet 2:1-3). While this is generally true, there were occasional exceptions (e.g., 1 Kgs 13:11-32). In one instance, a true prophecy came, however unwittingly, from an archenemy of Jesus Christ:

> One of them, Caiaphas, being high priest that year, said to them, "You know nothing at all, nor do you consider that it is expedient for us that one man should die for the people, and not that the

---

[7] James L. Crenshaw, *Prophetic Conflict: Its Effect upon Israelite Religion*, ed. Georg Fohrer, Beihefte zur Zeitschrift für die alttestamentliche Wissenschaft 124 (Berlin: de Gruyter, 1971), 49-61.

whole nation should perish." Now this he did not say on his own authority; but being high priest that year he prophesied that Jesus would die for the nation (John 11:49–51).

This is a true prophecy from the New Testament's perspective. Therefore, following Crenshaw's investigations, wherever a biblical prophecy came from, through whomever it came, in whatever circumstances it was given, it might be a true prophecy if its predictions happened, and if it is in harmony with the rest of the Bible.

Testing biblical prophecy appears to anticipate what is currently known as the scientific method. However, it must be remembered that prophecies are often more nuanced because they involve people. Therefore there can still be "grey areas." For example, was Jonah a false prophet because his prediction of doom for Nineveh did not happen (Jonah 3:4,10) or was he a true prophet because his preaching led to the repentance of the Ninevites (Jonah 3:5–9)? Was Huldah a true or false prophet because one of her predictions was correct (2 Kgs 22:19) and one incorrect (2 Kgs 22:20)? Was her score of 50% a "pass mark" or not? In turn, scientific predictions are more precise, more exact than prophecies, given that they deal with objects and physical forces, not persons.

The Scientific Method

The advent of empirical science—also called modern science or experimental science that uses the scientific method—certainly was one of the greatest leaps forward for the human race. It was a complete change—what Thomas Kuhn called a "revolution" or a "paradigm shift"—from what went before it, Aristotelian science. Modern science or empirical science is based on, concerned with, and verifiable by observation or experience rather than theory or pure logic. It is the practice of basing ideas and theories on testing and experience, capable of being verified or disproved by observation or experiment.[8]

---

8   Cf. Ian Hacking, "Introductory Essay," in Thomas Samuel Kuhn, *The Structure of Scientific Revolutions* (Chicago: University of Chicago Press, 1962), xiii. See

The technique for verifying theories and hypotheses, the scientific method, had been used *ad hoc* for about two hundred years before it was formalised by Francis Bacon in 1620, earning him the title "Father of Experimental Philosophy."[9] Bacon broke with Aristotle's philosophy, theology, and science, and its resurgence in scholasticism and the renaissance. Before attending Cambridge University, Bacon was educated at home by a private tutor, the Puritan John Walsall, who contributed to Bacon's Christian beliefs and "his distaste for what he termed 'unfruitful' Aristotelian philosophy, favouring instead the conviction that the human mind is fitted for knowledge of nature and must derive it from observation, not from abstract reasoning."[10]

In his *Novum Organum* (New or True Directions Concerning the Interpretation of Nature), Bacon detailed a new system of logic that he believed to be superior to Aristotle's old deductive and syllogistic approach. This is known as the Baconian method, precursor to the scientific method based on induction. The title of his dissertation is a reference to Aristotle's work *Organon*, which was the latter's treatise on logic and syllogism, the basis for his science, his natural philosophy. The front cover of *Novum Organum* cited Dan 12:4 which includes the words: "And knowledge shall increase!" In this light, using the scientific method, the early modern scientists went looking for and found God's laws of nature. About this endeavour Johannes Kepler said: "Science is the process of thinking God's thoughts after Him."[11] Auguste Comte called this early era—the fifteenth, sixteenth, and seventeenth centuries—the "theological phase" of modern science.[12]

---

	also https://www.merriam-webster.com/dictionary/empiricism (accessed 23 November 2020).
9	Peter Urbach, *Francis Bacon's Philosophy of Science: An Account and a Reappraisal* (La Salle: Open Court, 1987), 192.
10	Francis Bacon, *Of the Proficiency and the Advancement of Learning, Divine and Human* (1605) http://www.gutenberg.org/ebooks/5500 (accessed 22 November 2020).
11	Johannes Kepler Quotes, https://www.pinterest.co.uk/pin/811773901558228997/ (accessed 23 November 2020).
12	Hacking, "Introductory Essay," xxxiv.

The Scientific Method

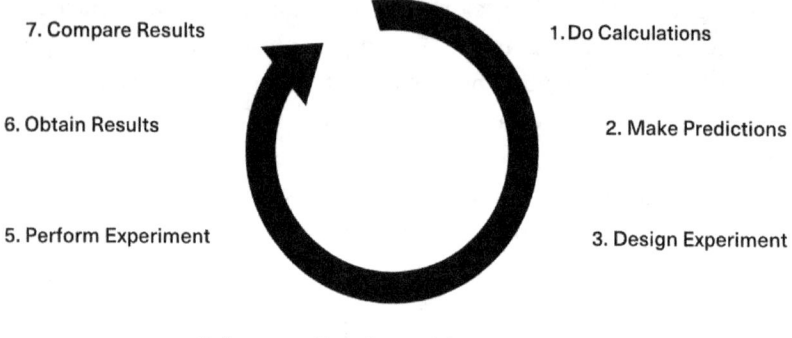

Figure 1

The scientific method may be called a "Reality Test." Figure 1 depicts it in more detail. The essence of the scientific method—what Robert Nola and Howard Sankey call its "meta-method" or its "meta-methodology"—is to find what a scientific theory or hypothesis predicts, and then to perform experiments and observe whether the prediction happens or not, i.e., if the theory's prediction is correct.[13] If it is not, the theory is rejected as wrong. It may be rejected altogether, or it may be modified in one or more ways, yielding a new theory which gives new predictions, which can then be tested against the scientific method, and so on. A scientific hypothesis must be falsifiable, implying that it is possible to identify a potential outcome of an experiment or observation that could conflict with predictions deduced from the hypothesis; otherwise, the hypothesis cannot be meaningfully tested.

Some disciplines require slight modifications of, or additions to, the general scientific method. When testing astronomical phenomena,

---

13   Robert Nola and Howard Sankey, *Theories of Scientific Method: An Introduction* (Montreal: McGill-Queen's University Press, 2007), 1.

for example, the scientific method relies on observations only, since we cannot perform experiments on stars, galaxies, comets, etc. When used with human subjects, such as in testing psychological theories and the efficacy of newly developed medications, the "double blind" technique is added to the scientific method in order to eliminate human expectations (the "placebo effect") as much as possible.

Testing scientific theories with the scientific method has exactly the same "meta-method" or "meta-methodology" as the test of prophets and prophecies; it is a matter of seeing what they predict and then of checking the prediction in the real world by observations and experiments. Because they both follow the same meta-methodology, testing prophecies and scientific ideas have common grounds. In particular, just as there are four steps of faith in testing prophecies, so there are four steps of faith in using the scientific method. We have noted one common ground already—both prophecies and scientific theories need to be falsifiable in order to be meaningful and able to be tested.

Faith in the Scientific Method

As with the test of prophets and prophecies, we need to have faith in the scientific method before we use it. Most scientists use it simply because it was passed on to them as how to do what they need to do to get their research done and publish their results. For those who have thought more about it:

> Scientists use the scientific method because it is evidence-based, standardized and objective in conducting experiments. The scientific method allows scientists to stick to facts and to avoid the influence of preconceived notions and personal biases in research processes, improving the credibility of research findings ... The scientific method involves a rigorous methodology that is aimed at minimizing prejudice.[14]

---

14   "Why Do Scientists Use the Scientific Method?" https://www.reference.com/science/scientists-use-scientific-method-887b9796714e7261 (accessed 23 November 2020).

These are excellent reasons for adopting and using the scientific method, but they are not proof of its validity. Some scholars such as Paul Feyerabend reject its comprehensiveness.[15] In turn, Nola and Sankey defend "the idea that there is such a thing as scientific method," and seek to justify, warrant, and legitimise it.[16] That it needs to be argued thus shows it is neither a fact nor self-evident but an article of faith. No experiments have been performed to verify the scientific method itself—you cannot use the scientific method to validate the scientific method. You either believe it is valid or you believe it is not valid.

To be an empirical scientist, one must believe the method is legitimate. The reasons for accepting and using it are irrelevant. One may believe it because it is in harmony with the Bible (Deut 18:21-22) as the present writer does. One may believe it because that has been the tradition of the scientific establishment for over five hundred years. One may believe it given that contemporary experts promote it. Moreover, one may believe it for any other reason/s. But, to be an empirical scientist, one must believe that it is valid.

Faith in the Results of the Scientific Method

Popper refined the theory of using the scientific method, showing that its purpose is not to verify hypotheses and theories, but to falsify them.[17] His arguments have won the day, with virtually everyone agreeing with him.[18] In order to verify that any theory or hypothesis actually is a law of nature—i.e., to know it is true and factual everywhere all the time—we would have to articulate it precisely and correctly, test it with infinitely accurate instruments, at every place throughout the universe, and at every time throughout the universe, past, present, and future, which, of course, is impossible on all counts. Therefore, human beings can never verify or determine conclusively whether a scientific

15 Paul Feyerabend, *Against Method*, 3rd edition (London: Verso, 2002), 23.
16 Nola and Sankey, *Theories of Scientific Method*, 1.
17 Popper, *The Logic of Scientific Discoveries*, 32-34, 40.
18 Martyn Shuttleworth and Lyndsay T. Wilson: "Falsifiability: Karl Popper's Basic Scientific Principle" https://explorable.com/falsifiability (accessed 23 November 2020).

hypothesis is correct, even when using the scientific method. Our experiments are always limited, never comprehensive enough, never extensive enough, and our instruments are never precise enough—they are never perfectly or infinitely accurate.

The reason why the scientific method is only capable of the falsification and never the verification of scientific theories is that, like the test of a false prophet, it is circular reasoning, as seen in Figure 1. It starts with the prediction of a scientific theory (or a prophecy) and ends with comparing the results of the experiment with the prediction with which you started. Circular reasoning can only prove whether the original proposition, theory, or hypothesis is consistent or not. If it is inconsistent, i.e., if its prediction does not happen, then it is false, and should be rejected. If the prediction does happen, then the theory, hypothesis, or proposition is consistent, but there is no way to tell by using the scientific method whether it is consistently right or consistently wrong.

This lack of certainty goes by the name of underdetermination:

> In the philosophy of science, underdetermination or "the underdetermination of theory by data" is the idea that evidence available to us at a given time may be insufficient to determine what beliefs we should hold in response to it. Underdetermination says that all evidence necessarily underdetermines any scientific theory.[19]

Said otherwise, in order to verify that a theory is correct, it would need to be tested with infinitely accurate instruments, at every place in the universe, at every time in the universe. Only if that is achieved can a theory claim to be verified. Since this is never the case, there is no proof that any scientific theory or hypothesis is true throughout the universe.

Similarly, giving an accurate prediction does not prove a scientific theory correct; it might still be a wrong theory with a lucky guess. The phlogiston theory of combustion—that flammable materials contain a substance called phlogiston that leaves it during combustion,

---

19   Kyle Stanford, "Underdetermination of Scientific Theory," in *The Stanford Encyclopedia of of Philosophy*, ed. Edward N. Zalta (Stanford University Press, 2021).

leaving ash of lesser mass—gave many correct predictions, for just over a century, e.g., burning wood, paper, candles, etc. That meant it was consistent. However, it was consistently wrong, as was later demonstrated. It was eventually proved wrong with the example of burning magnesium, whose ash, magnesium oxide, had more mass than the original magnesium. That means a step of faith is required to accept a theory as one of the laws of nature. A scientist could put it as follows: "I know this is only circular reasoning. I know it is only evidence for its correctness, not proof of it. But I have enough evidence, from this and other experiments, and from other considerations as well. Therefore, I am convinced it is right. Consequently, I will believe it is accurate—I will take a step of faith and act on it, basing all my future scientific theories and research on it." Incidentally, Kuhn used the word "conversion" to describe a scientist's changing from one scientific paradigm to a different one.

Unfortunately, what is "enough" evidence to be convincing is different for everyone. It is for this reason that certain scientists are convinced of theories by the available evidence, while others are not. Some scientists are convinced on a small amount of evidence, long before other scientists are convinced. For example, many scientists in the early twentieth century died still believing in classical gravity and classical mechanics. They claimed not having sufficient evidence to abandon classical mechanics and classical gravity and convert to quantum mechanics and Albert Einstein's theories of relativity. They died believing that one day refinements to classical mechanics and classical gravity would be found that explained everything satisfactorily.

In the case of the physical sciences, there is no inspired, infallible, inerrant book (or anything else) to test theories and hypotheses against. They always remain theories, never to be adequately and fully verified. Because there is no proof available, some wrong theories may go for years, even centuries, before being proved wrong, as the phlogiston theory was. As another example, before Einstein, the classical theory of gravity, or Galileo's theory of relativity, was published in 1632 in his *Dialogue Concerning the Two Chief World Systems*. Here is a summary of the Galilean theory:

> Galilean transformations, also called Newtonian transformations, [which are a] set of equations in classical physics that relate the space and time coordinates of two systems moving at a constant velocity relative to each other. Adequate to describe phenomena at speeds much smaller than the speed of light, Galilean transformations formally express the ideas that space and time are absolute; that length, time, and mass are independent of the relative motion of the observer; and that the speed of light depends upon the relative motion of the observer. Compare Lorentz transformations.[20]

Lorentz transformations are used in Einstein's theories of relativity that treats length, time, and mass not as absolute, but as dependent on the motion of the observer.

For over two hundred and fifty years, Galilean theory was considered consistent and made correct predictions. However, it was still wrong, "consistently wrong," and was eventually proved wrong in 1887. Its predictions were only "correct within experimental error" through that quarter of a millennium. In 1887, Albert Michelson and Edward Morley developed an incredibly accurate interferometer that showed a prediction of Galileo's theory of gravity was inaccurate.[21] So far, its replacements—Einstein's Special and General Theories of Relativity—have lasted for over a hundred years without any wrong predictions. Will they ever be proved wrong? We do not know. All we can say is that, so far, they have always given correct predictions within the parameters of our current scientific instruments. We believe Einstein's theories of relativity are correct, and we base the rest of our science on them at present.

---

20 https://www.britannica.com/science/Galilean-transformations (accessed 16 April 2022). See also https://www.britannica.com/summary/Galileos-Achievements (accessed 16 April 2022).
21 "Michelson-Morley Experiment" in https://www.britannica.com/science/Michelson-Morley-experiment (accessed 3 March 2022).

Incidentally, a negative result (an outcome contrary to the prediction of the theory under investigation) only demonstrates that something in the circle of the scientific method is wrong. It could be in the mathematical computations; it could be in the design, construction, or malfunction of the apparatus; or because of some contamination. However, with due diligence, including peer review, constant checking and rechecking, such errors are usually eliminated, so that it is only the consistency of the theory with reality that determines the results of the experiment.

One corollary of this analysis is that there are no such things as religious facts. The teachings of all the religions and their prophets are accepted on faith. Correct predictions do not prove that prophets are genuine—they could still be false prophets with a lucky guess. Many adherents will have strong faith in their religion's founder/s and their teachings, treat them as facts, and base their lives on them. They will accept the testimony of eye-witnesses as truthful statements of what happened and what was said, such as seeing, hearing, and eating with the risen Christ. However, we live by faith (Hab 2:4; Rom 1:17; Heb 10:38).

In exactly the same way, there are no scientific facts—no scientific theory can ever claim to be proved right, or determined, or established as a fact either. Any theory could be falsified by new experiments and new observations with more accurate instruments at any time. Claims that correct predictions concerning the cosmic microwave background radiation prove that the Big Bang theory is right or factual are mistaken. The correct predictions are evidence for the theory's correctness, but not proof of it. There is always a step of faith made. The more evidence we have, the smaller the step of faith needed—but there is always a step of faith required—it is never completely eliminated. We never know if three or three thousand years later a more accurate experiment will prove it wrong. Scientists, like Christians, live by faith and need to admit it.

To summarise, the second scientific step of faith is believing that a theory which has been tested using the scientific method and given correct predictions is not consistently wrong but consistently right, an accurate description of reality, and then acting on it. The more evidence we have, the smaller the step of faith made, but faith is always required.

Faith in the Extent of the Scientific Method

Suppose a group of scientists perform an experiment to find how the forces exerted by two electrically charged objects on each other depends on their distance of separation. They perform this experiment in Brisbane, Australia, at 10:00am on Thursday 10 February 2022, and get the result that it is inversely proportional to the square of the distance between them. Strictly speaking, all they have demonstrated is that at their location in Brisbane at 10:00am on Thursday 10 February 2022 the force exerted by charged particles on each other was proportional to the inverse square of the distance between them. Why should anyone believe that it is the same anywhere else in the universe or at any other time throughout the history of the universe?

    Someone might object: "We don't just believe it! We know it is true throughout all space—i.e., throughout the whole universe—and throughout all time—past, present, and future—because thousands of scientists and thousands of science students have performed similar experiments right round the world for hundreds of years and all got the same answer! No faith is needed!" We know they did. I have performed some of those experiments myself. Most likely, you have too. But how do we know we did not miss a time or a place or times and places where it was otherwise? How do we know if a law of nature is being broken now near Alpha Centauri, so we will not find out about it for over four years (at the speed of light)? Empirical scientists believe that all the laws of nature are uniform throughout the universe. The limited evidence that we have gives solid pointers in that direction, so it is a step of faith, not a leap of faith. But it is still a step of faith to believe that the universe is entirely regular.

    Recently, postmodernists have claimed that there are no such things as "universal truths"—true for everyone, everywhere, and at every time. They are at most only "true for you," but may not be true for anyone else, let alone for everyone else, everywhere else and "everywhen" else. Empirical scientists disagree with postmodernists on this point. Empirical scientists or modern scientists hold that the laws of nature are universally true—they operate infallibly throughout the whole universe throughout the whole history of the universe, whether people believe they do or not. Again, why one would believe this is true? One might believe it because the universe was created by a God of law and order. One might believe it because that is the tradition of empirical science for the past five hundred years. One, again, might believe it for any number of other reasons. But belief is crucial in order to be an empirical scientist.

The third article of the faith of empirical scientists is believing that the laws of nature hold true everywhere and "everywhen"—past, present, and future. The evidence from observations by human beings over the past five hundred years or so—an extremely tiny proportion of the entire history of the universe—points in that direction, so it is a step of faith, not a leap of faith. But faith it remains; it has not been proved or verified.

Faith in the Scientists Who Use the Scientific Method

We need to have faith in scientists that they will honestly report the results of their experiments and what those experiments indicate. Someone might object that the scientific method involves only facts and therefore we do not need any faith in those performing the experiments. Anyone at any time and in any place can repeat these experiments for themselves to check and see that the results obtained are genuine. Similarly, anyone can check their theory, their equations, their apparatus, and their instruments for themselves as well.

While this sounds good in theory, does it really work out that way in practice? Where would I, or any other scientist for that matter, obtain the multibillion dollars necessary to build and launch a telescope into space for ourselves ($1,000,000,000 to build and launch in 1995, plus $100,000,000 per year to operate) to check the images reportedly obtained from the Hubble Space Telescope? Failing that, would I be allowed to launch on the next space mission to check if the images that are claimed to have come from the Hubble Space Telescope really did? Finally, who will pay for me or anyone else to spend twenty years or more at a university to learn the theory behind the experiments and the experimental equipment?

The answer is, of course, in the negative to these and similar questions. We need to have faith in the scientists who run the Hubble Space Telescope and all the other pieces of very expensive scientific apparatus and instrumentation. We need to have faith in the scientists that they are competent in the use of the equipment, and that they are honestly reporting their results. Actually, we need to have faith in them at all seven steps in the scientific method, as shown in Figure 1.

Occasionally, scientists will include some kind of certification or perhaps a statutory declaration that they have done all of this. Even if they do not, it is the tacit assumption that they have. Regrettably, this has not always been the case. There have been occasional examples of deception. Piltdown Man was fabricated from a modern human skull, some chimpanzee teeth, and an orangutan jaw, and then "doctored" to appear millions of years old, and thus made to look like a missing link in human evolution. To mark April Fools' Day, National Geographic News summarised some historic scientific hoaxes: Piltdown Man, Cardiff Giant, Archaeoraptor, and Bigfoot. "Not only was the Piltdown skull itself fraudulent but the entire mammalian fauna of the gravels had been planted and the human artefacts manufactured."[22]

In religion, there have been cases of "fake miracles," like instances of bleeding statues being reported to bolster people's faith in the god/s and/or goddess/es of that religion. For example, the *Irish Times* reported on one such hoax in 1920 with the headline: "The 'Templemore Miracles': How a fake bleeding statue led to an IRA truce."[23] The apocryphal additions to the biblical book of Daniel, *Daniel, Bel, and the*

---

22 L. B. Halstead, "New Light on the Piltdown Hoax?" *Nature* 276:5683 (1978): 11-13.
23 https://www.irishtimes.com/culture/heritage/the-templemore-miracles-how-a-fake-bleeding-statue-led-to-an-ira-truce-1.4328392 (accessed 1 March 2022).

Dragon, tell the story of how Daniel exposed another such fraud (Dan 14). The Bible warns against such tricksters multiple times (e.g., Deut 13:1–5; 18:20–22; Matt 24:24; Mark 13:22; Rev 19:20). Hence the Bible's tests of false prophets to protect believers from them.

Incidentally, I do trust or have faith in the honesty of scientists at the CERN collider, etc., unless the contrary is proved beyond reasonable doubt. In the same way, I trust or have faith in Matthew, Mark, Luke, John, Moses, Isaiah, Jeremiah, Ezekiel, and all the other Bible authors that they have honestly reported what they saw, heard, and experienced. To this end, several biblical authors give a certification—the first century equivalent of a statutory declaration—that they have done this truthfully. For example, the Apostle John wrote:

> This is the disciple who testifies of these things, and wrote these things; and we know that his testimony is true. And there are also many other things that Jesus did, which if they were written one by one, I suppose that even the world itself could not contain the books that would be written. Amen (John 21:24–25).

John is testifying that he is telling the truth and nothing but the truth. However, he tells us that it is not the whole truth because there was simply too much to report. He has given us a "typical sample" of what he saw and heard Jesus do and say. We could see also John 20:30–31, 1 John 1:1–3, and Luke's attestation in Luke 1:1–4. It is the tacit assumption that all other biblical authors are doing likewise. The Bible then offers for everyone to "repeat the experiment"—to believe in or accept Jesus Christ, and experience this for themselves, to "taste and see that the Lord is good; blessed is the man who trusts in Him!" (Ps 34:8).

With all of that being said, the truth or otherwise of a scientific theory does not depend on the honesty or anything else about the scientists who propose and/or promote it. In the case of Gregor Mendel who discovered and enunciated the laws of heredity (about dominant and recessive genes, etc.), he dishonestly reported the findings of his observations and experiments to help convince his peers of the accuracy of his theory. The combination of parents' genes in each of their offspring at reproduction is a random process, having an average or mean, and a spread measured by its standard deviation. Recent observations and more accurate measurements of experiments on pea plant reproduction indicate that Mendel "fiddled" or "cooked" his results to make his predictions look more obvious and more accurate. As Michael Starboard notes, "the number of experiments in which Mendel's data were very close to expectation was too great to be believed."[24] This also demonstrates the importance of truthfully calculating, including, and reporting the experimental error and standard deviation in the results of experiments. However, despite his dishonesty, Mendel's theory of dominant and recessive genes has so far stood the test of time and further, more accurate experimentation.

Induction

Finally, a word needs to be said on where scientists get the ideas upon which, and from which, they develop their scientific hypotheses and theories. Following Bacon, the belief was once held that scientific hypotheses must be generated via induction—by performing many experiments, usually drawing graphs of the measurements taken, looking for patterns in the data, and then, from all that data, inducing the relationship/s between the variables. However, Popper showed that one can get a scientific hypothesis from anywhere, not just via induction.[25] Popper called this initial conceiving of a theory its "psychological" stage—it originated in the *psyche* or mind of the scientists.[26] What makes it scientific is not its origin, but the criterion for its acceptance or rejection—the tests and observations made using the scientific method. Feyerabend also rebelled, this time correctly, against the forbidding of what he called "ad hoc hypotheses."[27] Much of Feyerabend's observed

---

24   Michael Starboard, "Did Famous Genetic Scientist Gregor Mendel Fake His Data?" https://www.thegreatcoursesdaily.com/gregor-mendel-fake-data/ (accessed 1 March 2022).
25   Popper, *The Logic of Scientific Discoveries*, 27–32.
26   Popper, *The Logic of Scientific Discoveries*, 30–31.
27   Paul Feyerabend, *Against Method*, 4th edition (London: Verso, 2010), 8.

"scientific anarchy" is in the multiplicity of ways scientists have conceived of, derived, and developed hypotheses, some very creative and some very unconventional. One of his examples was showing that Galileo did not, and indeed could not obtain the heliocentric model of our solar system by induction, but by what he calls "counterinduction," which denotes "thinking outside the box."[28] Like prophecies, scientific hypotheses may be drawn from anywhere.

The observation of the present writer is that induction played a much more significant role earlier on in the scientific investigation of all the different phenomena. Hacking and Kuhn observed that, before one single paradigm emerges as supreme in any branch of the physical sciences, "we have a pre-paradigm period of speculation ... there was simply no way to sort things out, no set of agreed problems to work on, precisely because there was no paradigm."[29] In that atmosphere, scientists did a lot of experimentation, analysing results, plotting graphs, and trying to recognise any patterns upon which to induce a theory or hypothesis. Any hypotheses generated would then be tested using the scientific method.

Nowadays, new phenomena are rarely examined from scratch. There are well established theories with all their equations and past experiments in all branches of the physical sciences. What are currently tested in experiments are new implications drawn from what those theories predict under different conditions. The Higgs Boson, for example, was discovered because theory predicted it, experiments were designed and performed accordingly, and it was eventually identified. No one did multiple high energy experiments in the Large Hadron Collider, examining particles that were produced, and then using induction on the results.[30] Many other techniques have been employed to generate theories, such as purely theoretical considerations, dimen-

---

28 Feyerabend, *Against Method* (3rd edition), 116.
29 Hacking, "Introductory Essay," xxv.
30 "New Results Indicate that New Particle Is a Higgs Boson" (2013) https://home.web.cern.ch/news/news/physics/new-results-indicate-new-particle-higgs-boson (accessed 3 March 2022). See also "The Search for and Discovery of the Higgs Boson" https://en.wikipedia.org/wiki/Higgs_boson (accessed 23 November 2020).

sional analysis, and from parallels drawn with theories from other fields of science.

Therefore, scientific hypotheses can be drawn from anywhere—from induction, by modifying a previously falsified theory, from dimensional analysis, from a séance, from the Bible, or from "sudden flashes of inspiration," as Popper has shown. The most bizarre example I know is the determination of the chemical structure of the benzene molecule [C_6H_6]. Friedrich Kekule and Johann Loschmidt received the idea of the benzene molecule being a flat ring, not a chain, in a dream (of a snake biting its tail) or a nightmare (where carbon atoms danced around poking fun in "A Ring A Ring A Rosy").[31] Nevertheless, such theories and hypotheses can qualify as scientific if they are then tested by the scientific method and are shown to give correct predictions continuously, as the structure of the benzene molecule has.[32] That is why, in Figure 1, above, the origin of a theory or hypothesis, Point #0, is outside the actual circle of the scientific method.

The most that can be said is that the origin of the idea for a theory might make one suspicious of it, but it does not prove the theory wrong or unscientific. Neither does the dishonesty of the scientist/s reporting on experiments to check a theory's predictions prove that theory wrong or unscientific. At most, it might make one suspicious of them. Therefore, the fact that a theory is derived from the Judeo-Christian Bible, the Qu'ran, a séance, the Bhagavad Gita, a nightmare, etc., does not preclude it from scientific consideration. It might make some scientists suspicious of them, but that is all. In the same way, the dishonesty of some evolutionists does not prove the theory of evolution wrong. At most, it might make scientists suspicious of it, but that is all. The only thing that can prove theories and hypotheses wrong is observations and/or experiments proving their predictions to be wrong.

---

31   https://www.britannica.com/science/benzene (accessed 23 November 2020).
32   Popper, *The Logic of Scientific Discoveries*, 32–34.

Conclusion

This article has argued against the claim of atheists that "science is founded on facts, religion is founded on faith," and that they are in irreconcilable conflict with each other. It sought to show that the empirical sciences involve faith, particularly faith in and around the scientific method.

This article showed that the scientific method is not antagonistic to the Christian Bible, but in harmony with it. The agreement or harmony was established by drawing a parallel with Deuteronomy 18:21-22, where the test of false prophets and false prophecies depicted there has the same meta-method, the same meta-methodology as the test of false scientists and false scientific theories, what we call the scientific method.

It proceeded to show that the scientific method itself is an article of faith—its validity cannot be proved logically or scientifically, especially not by the scientific method itself. We can amass evidence and arguments for its adoption, but, in the end, we either believe it is legitimate or we believe it is not.

It demonstrated that the scientific method is circular reasoning, and, being circular reasoning, can only prove a theory or hypothesis consistent or inconsistent. If it is proved inconsistent—i.e., its prediction does not happen—then that theory is discarded as false. If its prediction happens—i.e., the theory is consistent—the scientific method cannot tell if it is consistently right or consistently wrong. In Popper's words, the scientific method is only a means of falsification. A step of faith is then needed to believe that the theory is true. The more evidence we have, the smaller the step of faith required, but this step of faith is never eliminated.

The scientific method itself cannot predict if the same experiment done in a different location or at a different time will always yield the same result/s for entities deemed to be "universal constants," such as G, the gravitational constant, c the speed of light, ε_0 the permittivity of free space, μ_0 the permeability of free space, etc. So far experiments always have assumed universal validity. But, to generalise further, we need to believe that the laws of the universe are the same at all places throughout the universe, and at all times throughout its history—past, present, and future.

While in theory anyone should be able to repeat any scientific experiment and confirm its results for themselves, in practice this is not always feasible, mainly due to the complexity in the design and manufacture of the necessary equipment, and its prohibitively high cost. Therefore, we also need to trust or have faith in the scientists who perform and report any experiments, that they have honestly recounted their findings, experimental error, etc.

The theory to be tested by the scientific method can originate from anyone, anyhow, in any way, in any circumstances, at any time. It may be derived from induction. It may be derived from dimensional analysis. It may be drawn from the Bible. It may be totally concocted by the scientist/s. What makes it scientific is whether it can be and has been tested using the scientific method, not its origin.

Just as there are no religious facts, there are no such things as scientific facts either. Any scientific theory, no matter how confidently it is believed, by no matter how many prominent scientists, for no matter how long, with no matter how many correct predictions made so far, may be proved wrong at any time by a future, more accurate experiment, just as classical gravity was proved wrong after some two hundred and fifty years of making correct predictions.

Therefore, it is not the case that: "Religion is founded on faith and science is founded on facts." Instead, both religion and empirical science are founded on both facts and faith. Scientists, just like Christians, live by faith in the ways enumerated and discussed above.

The author reports there are no competing interests to declare.
Received: 17/02/22 Accepted: 18/07/22 Published: 18/09/22

Bruce Craven: Contrarian or Questioning Thinker?

John Pilbrow

Abstract: This article continues the author's tribute to Bruce Craven, published on the ISCAST website earlier this year and reproduced here, revised and expanded, in the Appendix. Craven's relevant contributions are reviewed in the hope that both ISCAST members and other readers can appreciate his robust thinking at the nexus of Christianity and science. The approach is straightforward, the author focusing on Craven's articles published in *Christian Perspectives on Science and Technology*, where he gleans true gems and a few weaknesses. What emerges at the end of this exploration is the portrait of Bruce Craven as a Christian "questioning thinker" who—equipped with the specific skills of his mathematical expertise—is able to inspire his readers today, as he did in the past.

Keywords: Bruce Craven; creation narratives; divine presence; evolution; scientific method

In his writings as well as during ISCAST meetings and conferences, Bruce Craven always came across as a somewhat *contrarian* thinker because of the kinds of questions he posed. What is attempted here is a review of some of his thought as exemplified in a number of articles

---

Emeritus Professor John Pilbrow is a Life Fellow and former President of ISCAST.

available on the ISCAST journal's website.[1] On balance, what emerges is not so much a *contrarian* thinker, but rather someone who wanted more *i*s dotted and more *t*s crossed, in fact, very much a *questioning* thinker.

We now consider some of the issues that Bruce raised and on which he pondered in some depth, and see what we can learn from them. This discussion does not exhaust all that Bruce Craven wrote or thought, but should serve to illustrate why we are much in his debt.

For those readers who did not know Bruce or who were unaware of his contributions to ISCAST, biographical information is provided in the Appendix.

God's Involvement in the World

When thinking about science from a Christian perspective, Bruce observed that to state that "it is all 'God's world' can become a meaningless platitude if our system excludes God from any continuing role in the world." Indeed, he continued to try to understand how God is involved in the cosmos, stopping short of wanting to put God's name in scientific papers. Rather, the question of God's involvement[2] comes in at a philosophical and theological level, where one can think about purpose.

---

[1] Go to http://journal.iscast.org/ (search for *Craven*). Articles on the website of *Christian Perspectives on Science and Technology*: "Editorial" (with John Pilbrow) vol. 10 (December 2014); "Working Hypotheses in Science" vol. 8 (January 2012); "What Doubt Is Reasonable?" vol. 7 (December 2011); Review of Michael Poole's book *The New Atheism: 10 Arguments that Don't Hold Water* vol. 6 (May 2010); "How Useful is Unpredictability? A Mathematician's Thoughts on Gambling" vol. 6 (April 2010); "Evolution—A Short Guide for the Perplexed" vol. 4 (October 2008); "What Does Genesis Tell Us?" vol. 4 (June 2008); "Explanation and Belief" vol. 4 (April 2008); "Ethics in Research" vol. 2 (December 2006); "Are God's Actions Hidden in Chaos?" vol. 2 (June 2006); "Death of Science?" vol. 1 (November 2003).

[2] The topic was discussed in six volumes resulting from a series of conferences jointly organised by the Center for Theology & the Natural Sciences, Berkeley, and the Vatican Observatory. Under the general heading, *Scientific Perspectives on Divine Action*, here are the titles of the six volumes in this series: *Quantum Cosmology and the Laws of Nature*; *Chaos and Complexity*; *Evolution and Molecular Biology*; *Neuroscience and The Person*; *Quantum Mechanics*; *Twenty Years of Challenge and Progress*. ISCAST members will be interested to know that the last chapter in the sixth and final volume was written by ISCAST Fellow, Mark Worthing.

He wondered whether Chaos Theory might help us understand how God acts. He noted that many physical systems are extremely sensitive to initial conditions, so that a small unobserved input can produce large consequences later, and moreover can behave in a seemingly random way. He was not alone in wondering whether perhaps God intervenes[3] in His creation by such small inputs, without violating the regularities that we call physical laws. He thought the world may be less deterministic, and more open to the future, than many suppose.

What Does Genesis Tell Us?

Bruce readily acknowledged that what we know from modern science forces us to rethink how we understand the early chapters of Genesis. In fact this is an ongoing necessity since those chapters continue to be at the core of much public controversy. He had this to say about Genesis:

> A first reading of Genesis suggests a creation in six literal days (but what was a day before the sun was there?). Many early Christian writers did not understand it so literally. Calculations from lists of ancestors suggest about 6000 years since the creation (though only if we had complete records, but we don't). Ancient writers had not our technical terms, and often expressed ideas by stories. We must try to understand the main point of the story, but to insist on a literal interpretation of every detail does little to praise God. They, like we, were concerned with how things began; but they were interested in purpose—what was it for?—whereas we are much more interested in method—how did it come about? The two don't have to fight. The authors of Genesis shared common traditions with their neighbouring peoples. These included creation from chaos, in a number of stages. But Genesis understood it differently. Instead of a number of gods fighting in the sky, a random world, and humans as an afterthought (only to feed the gods), Genesis describes a world made by a single creator, a world with coherent structure, and humans as important, made with

---

3   It is probably better to speak in terms of God interacting with his Creation rather than intervening. The latter is open to the idea that God only acts occasionally, rather than upholding the universe continually.

something Godlike in them, and God feeds them. If this is what
Genesis 1 was telling its first hearers, then we need not get hung
up about days.[4]

This statement is consistent with the point of view expressed by the
late Dr John Thompson[5] and, in the light of that, Bruce rejected the
false idea that God made the world look like it was old, an argument
sometimes used by those promoting a literal view of Genesis 1-3. After
all, Genesis 1-3 is not the only scriptural story of creation. It is worth
pointing out that the 2017 ISCAST Lecturer, Tom McLeish, has noted
there are more than twenty Creation Narratives in Scripture.[6]

In passing, while on the subject of the early chapters of Genesis,
Bruce pointed out that while the New Testament does not mention research,
to fulfil the requirements of Gen 1:28 necessitated research and
observation, which we know to be the key to modern science.

Explanation and Belief

Bruce noted that scientific explanations depend so often on analogy
with simpler things and he questioned: what were the limits of this
approach? He readily acknowledged that the scientific enterprise ever
seeks to move closer to the truth. He made a particularly interesting
observation that a clear explanation in one culture may be incomprehensible
in another. Accordingly, he pondered how we may choose between
different possible explanations. For example should we adopt
the simplest explanation (Occam's Razor)? Or, alternatively, choose

---

4   "What Does Genesis Tell Us?" *Christian Perspectives on Science and Technology* 4
    (June 2008) https://journal.iscast.org/past-issues/what-does-genesis-tell-us (accessed
    on 10 May 2022).
5   J. A. Thompson, *Genesis 1-3: Science? History? Theology?* (Melbourne: ISCAST and
    Acorn Press, 2007).
6   Tom McLeish, "Biblical Creation: Over 20 Creation Accounts in the Bible?"
    (28 October 2019) https://iscast.org/news/biblical-creation-over-20-creation-
    accounts-in-the-bible/ (accessed on 10 May 2022).

an explanation with a beautiful equation à la Paul Dirac?[7] This still involves making a judgment in a given case.

Further, he was well aware that there are, of course, different levels of explanation. For example, the otherwise discredited theory of epicycles to explain planetary motion nevertheless remains important in navigation. Then there are theories with predictive capacity, e.g., gravitation. He noted that Newton's great discovery was that the physics of the falling pebble and planetary motion involve the same theory.

Bruce frequently asked whether there is intelligence and/or purpose behind observed phenomena. He wondered what level of autonomy the universe possesses, something that Polkinghorne discussed in terms of the contrast between human free will and his free process defence concerning the intrinsic behaviour of the universe.[8]

Limits to Science

Bruce recognised that many individual scientists have not thought through the philosophy of science that they actually use and default to scientism, the idea that if something cannot be demonstrated scientifically it is not meaningful knowledge. This situation demonstrates the long reach of the Vienna Circle's Logical Positivism from the 1920s and the 1930s. I guess what he wanted, above all, was for all scientists to have thought deeply about their science and the basis on which that science rests.

Bruce was certainly aware that our present knowledge and current understanding will always be tentative, but he realised that does not excuse us from embracing the best understanding we can find. He was rightly concerned about the limits of science and in several places

---

[7] As an Honours student a long time ago, I came across Dirac's relativistic theory of the electron and the key equation which, still to my mind, is one of the most wonderful equations in the whole of science. It not only illuminated the relativistic behaviour of electrons, but predicted antimatter (negative electrons or positrons used in PET scans in medicine today). The prediction predated the discovery of antiparticles by several years.

[8] John Polkinghorne, *Science and Providence* (London: SPCK, 1989), 66.

referred to Nobel Laureate Sir Peter Medawar, a rationalist, who had this to say:

> There is no quicker way for a scientist to bring discredit upon himself and upon his profession than roundly to declare—particularly when no declaration of any kind is called for—that science knows or soon will know the answers to all questions worth asking, and that questions that do not admit a scientific answer are in some way non-questions or "pseudo-questions" that only simpletons ask and only the gullible profess to be able to answer.[9]

Though Bruce did not write a specific article to articulate a Christian understanding of the philosophy of science, we can glean something of how that would look from the points raised in this reflection. Bruce always wanted to have a distinct Christian perspective. It is in this respect that his understanding went beyond Medawar's statement, even though he found that to be a helpful insight.

Further, he was concerned about pressure placed on scientists in certain contexts to assert opinions not supported by the data. He referred particularly those in Christian Colleges (especially in the USA) who are not free to express an opinion on evolution except to dismiss it.[10]

Various Roles Played by Doubt

Here Bruce explores a range of issues that involve doubt both as a positive and as a negative influence.

Doubt often plays a significant role in science when seeking to judge between two or more competing theories. And there is also doubt that some have regarding the reality of God and the truth of the Christian story. But he noted an insidious kind of doubt resulting from research funded by large international enterprises that wanted scientific

---

9   Peter Medawar, *The Limits of Science* (Oxford University Press, 1984).
10   Bruce would have had in mind the situation faced by the 2008 ISCAST Lecturer, Richard Colling, author of *Random Designer* (Bourbonnais, IL: Browning Press, 2004), who ultimately resigned from a Christian college in the US after being prevented from teaching evolution there.

research to support their product. He refers to the exposé *Merchants of Doubt* that reported in detail on the attempt to sow the seeds of doubt on an unwary public and the science community.[11]

He also posed the question, "What is reasonable doubt?" particularly in the realm of potential catastrophes. He realised that decision makers cannot prevaricate forever.[12] Then he asked, "What counts as good scientific evidence?" He mentioned the conflict between Big Bang Cosmology and Steady State Theory that was settled eventually in 1964 in favour of the Big Bang after the observation of the microwave background from the early universe.

With regard to the role of prediction in science, Bruce referred to neutrinos that were predicted long before they were detected. He noted that dark matter is required in cosmology, but has not actually been identified as yet. Here I would add gravitational waves, predicted by Einstein's General Relativity in 1915, but not observed until a century later.[13]

Bruce also noted the increasing pressures on researchers. For example, universities have to some extent become businesses and without external funding (particularly from industry) some university research may not be possible. There are ethical issues involved.[14]

Evolution—A Short Guide for the Perplexed

Bruce remained perplexed about evolution, particularly because he recognised that Darwinism has often become a worldview that goes beyond the realm of biological evolution, and imparts to it a broad function and purpose that cannot be deduced from biology alone. He was,

---

11 Naomi Oreskes and Erik M. Conway, *Merchants of Doubt: How a Handful of Scientists Obscure the Truth on Issues from Tobacco to Global Warming* (Bloomsbury, 2010).
12 See *A Reckless God: Currents and Challenges in the Christian Conversations with Science*, ed. Roland Ashby et al. (Melbourne: ISCAST and Morning Star, 2018), 131, from the review of Sir John Houghton's autobiography, *In the Eye of the Storm*.
13 See Stephen Ames and John Pilbrow, "Gravitational Waves Discovery Opens New Way of Looking at the Universe" in *A Reckless God*, 286.
14 For more information regarding ethical issues in science, see Craven, "Ethics in Research" https://journal.iscast.org/past-issues/ethics-in-research (accessed on 10 May 2022).

however, well aware that the science of evolution is based on evidence from four kinds of observations:

(a) The earth is much older than 6000 years;
(b) Many species are known to be extinct;
(c) There are demonstrable common biological ancestries;[15]
(d) Neo-Darwinian natural selection operating locally (for closely related species).

While he questioned, "If we say evolution has been established, does that apply equally to (a)-(d)?" he did not address the technical issues in detail as a biologist or palaeontologist might have done, but rather was responding more as a mathematician, looking for a level of proof as one might in regard to a mathematical theorem. He was also unhappy with the sloppy use of "random" in much evolutionary discourse.[16] But, above all, Bruce wanted evolution to involve purpose. I respond by saying that to deal with such matters, we need to delineate the basis of science from wider philosophical and theological issues to do with meaning and purpose. Just as the presuppositions which underpin science are not themselves derivable within science, so any attempt to inject purpose into the discussion must be at the philosophical and theological level, but cannot be incorporated in the science itself.

He referred to the dispute between the late Stephen Jay Gould (Harvard) and Simon Conway Morris (Cambridge) during the late 1990s regarding what would happen if the evolutionary tape were rerun. He was encouraged by Conway Morris' arguments outlining the basis of evolutionary convergence. That is, there are islands of stability, random processes are involved, and not all outcomes are possible. This is the kind of language that Bruce as a mathematician would have understood very well. In fact this is probably about the nearest one could get to Bruce's quest for purpose in evolution.

---

15 Graeme Finlay, *Human Evolution* (Cambridge University Press, 2014). Finlay is an Evangelical Christian.
16 It is noted that Bruce's discussion of randomness in *Evolution—A Short Guide for the Perplexed* and in *Working Hypotheses in Science* are found to lack rigour.

Unpredictability

Bruce's interests and inquisitiveness knew no bounds. In thinking as a mathematician about unpredictability he wondered why it was that so many people want to gamble. He concluded that the reason is that as we are no longer hunter gatherers, our lives do not involve the same level of risk as experienced by earlier humans. An interesting observation.

The Scientific Enterprise

Bruce was concerned that science as we know it might not last. His brief account of the history of science, especially of modern science in Europe, may be contrasted with the short article by Peter Harrison.[17] Science emerged in a climate of opinion that nature is not capricious, a period ripe for technological inventions particularly in navigation.

While Bruce considered the death of science as not inevitable, nevertheless he thought the danger was real. This is his rather bleak assessment of the situation as he saw it.

> The scientific enterprise will not automatically continue in our changed social climate. If it is to carry on, some scientific leaders may have to put as much effort into influencing public opinion, as they do in raising funding. Scientists must show their concern about the use, or often misuse, of their knowledge. And some imagination is needed, on how to interest the younger generation in science.

It would have been interesting to see how Bruce would have recast this statement as a challenge for Christians working in the sciences.

---

17   ISCAST Fellow Peter Harrison, an acknowledged international scholar on the rise of modern science in Christian Europe, has written at length on the topic. We note a recent short article, "Christianity: The Womb of Western Science" (in *A Reckless God*, 17) that captures the essence of his thought.

Conclusion

What can we say by way of a summary of Bruce's ideas and thinking? Bruce was in some ways a contrarian thinker, not because he wanted to avoid having to decide the truth or otherwise of a major scientific proposition, but rather because he wanted the best basis to be able to judge for himself. In discussions following a variety of presentations at ISCAST events, Bruce would not let us simply accept something because someone had said it, but he always wanted us to be sure we understood what we had just heard. Perhaps he was something of a terrier—a deliberative thinking terrier. He was always right to demand proper attention to the basis on which major scientific conclusions were or are made. The mathematician in him sometimes looked for a deeper level of certainty than can be guaranteed in science. My comment is that the empirical sciences involve a more subtle assessment of theoretical understanding.

Knowing that Bruce struggled with the questions as to how God interacts with the world, it would have been interesting to know what he might have said about prayer, but this does not crop up in any of his articles published in *Christian Perspectives on Science and Technology*.

Bruce was saddened by those Christians who keep science and faith in separate boxes. He recognised that for such people to integrate their understanding of faith and science would involve rethinking their understanding of both and, ultimately, to be able to embrace and celebrate both.

I'm sorry I did not have opportunities to discuss more of these issues with Bruce in recent years. We need the Bruces of this world to hold us to account for the views we hold and to be prepared to modify them when it is obvious that becomes necessary. It is my hope that this reflection will help us all to understand better the kind of thinker that Bruce Craven was, not so much a *contrarian* thinker, but rather a *questioning* thinker. This should give us all something to think about! Bruce would not settle for glib answers or for superficial thinking. He always sought to challenge us to dig deep.

Appendix

Bruce Desmond Craven, 1931–2022[18]

We report with sadness the passing of Dr Bruce Craven, a long-time Fellow of ISCAST, on the evening of 25 January 2022, after a long illness.

Bruce, who was elected a Fellow of ISCAST in the early 1990s (and a Life Fellow in 2011), had participated in the former Victorian Research Scientists' Christian Fellowship (RSCF) from the late 1950s for about 20 years. When the ISCAST online journal, *Christian Perspectives on Science and Technology*, was established, Bruce was its Founding Editor. In addition to judicious reviewing of submitted articles, which sustained the ISCAST ethos, Bruce himself contributed across a wide spectrum of issues, including as indicated in n. 1 above.

An only child, Bruce was brought up in Hampton, a Melbourne suburb, and, apart from his time in the United Kingdom, lived in the same family home until he had to move into aged care some years ago. Anyone who ever visited his home would have seen the extensive bookcases in hall and rooms lined with mathematical journals and books on an endless variety of topics!

Bruce attended Hampton High School until he was awarded a Scholarship to Wesley College. At Wesley, he learned French as part of the curriculum, but he also took advantage of voluntary German lessons after school. This enabled him to read mathematical journal articles not only in English, but in French and German as well.

During the early 1950s, Bruce graduated with both BSc and MSc degrees with First-Class Honours in Mathematics at the University of Melbourne. In 1955 he spent a year working in industry in the United Kingdom, followed by several years as a Senior Research Physicist at Australian Paper Manufacturers Melbourne. During this time he gained a further degree from Melbourne University, BA (Hons) in Sta-

---

18 An extended biography may be found in my tribute to Bruce, at https://iscast.org/news/tribute-to-bruce-craven/ Some details presented here were obtained from the notification of Bruce's death to the Australian Mathematical Society, and are used with permission.

tistics, again with First-Class Honours. In 1962 Bruce was appointed Lecturer in Mathematics at Melbourne University, ultimately becoming Reader. He was awarded a well-deserved DSc in 1973.

During his academic career, Bruce also taught himself Russian, to the point where he was able to give lectures in Russian during visits to Moscow. Something of an adventurer, he once explained that during a visit to Moscow he decided to buy a bus ticket and traveled around the outer suburbs, something his Russian maths colleagues thought wasn't such a good idea for a foreigner. But that was Bruce, often somewhat unpredictable and yet unperturbed by apparent difficulty.

Bruce was in many ways, quite self-contained. He didn't indulge in small talk such as local gossip about football, cricket, or sports in general.

Bruce contributed much to the faith-science conversation here in Melbourne for more than five decades, for which many of us remain profoundly grateful. His faith in Christ was firm and informed, and he was a loyal member of the congregation at Brighton Church of Christ for most of his life. He was a good friend to ISCAST and we'll all miss his deep and insightful comments, some of which are explored above.

The author reports there are no competing interests to declare.
Received: 12/05/22 Accepted: 06/09/22 Published: 22/09/22

On Subjects, Objects, Transitional Fields, and Icons: The Semiotics of a New Paradigm in Human Studies

Marcello La Matina

Abstract: To save what is human (and humane) about the human sciences, the subject/object dyad must be abandoned in favour of a semiotic and an anthropological point of view. This viewpoint draws on the interaction of several signifiers in dialogue with a salient space similar in nature to the transitional field of psychoanalysis and—via an interpretation of that space—to the iconic function of human culture as seen by patristic wisdom. To attain this viewpoint entails abandoning the idea that the human sciences are supposed to *explain* the human being. Their task is to clarify the plural and ecological character of humans.[1]

Keywords: Anthropic zones; Byzantine icons; human sciences; ontology; person; semiotics; subject/object dyad; transitional objects

In all epistemologies, old and new, the subject/object dyad plays a crucial role. It is commonly believed that every genuine act of knowledge is oriented towards an object and, at the same time, explained as the doing of a subject. The same occurs with all human actions, since

---

Marcello La Matina is Professor of Semiotics and Philosophy of Language at the Department of Human Studies, University of Macerata, Italy. The author expresses his gratitude to the *CPOSAT* referees, whose comments contributed to bringing this article, which has known a long gestation, to the current form. He also acknowledges that he alone must be held responsible for any oversights and errors still to be found in the text.

knowing is the very mode of existence for living human beings. Yet, despite its pervasiveness, the subject/object binary is not that simple to read from an epistemological perspective. A theoretical storm about this topic has been gathering strength for decades, one that hinges on the current and future meaning of the humanities. Recently, the dispute seems to have reached the climax as a final showdown between the human and the natural sciences and, on a deeper level, between the cosmological vision of the human phenomenon and an anthropological vision of the *Umwelt*, or environment.[2] In a sense, the way is open from a human to a nonhuman ontology, passing through an object-oriented ontology.[3]

It is my conviction that semiotics, one the one hand, and patristic wisdom, on the other, can make an important contribution to this debate. Here, semiotics denotes the study of sense and signification, while

1   Although the term "human sciences" is widely used and accepted, a unified, reasoned definition that defines the field in a way that is acceptable to all is missing. Sometimes, human sciences are defined in opposition to the natural sciences; at other times, they are associated with the latter and differentiated only in relation to the role of the analysing subject. In this article, I seek to define the field in the sense of the Latin locution *studia humana*, which includes the relationship between human studies and concrete human beings.

2   The attempt to reconcile the anthropic and the cosmic perspectives could benefit from reference to the worldview of the early Christians, perhaps by comparing their sense of the cosmos with the cosmology emerging from quantum physics. Excellent work in this regard has been done by Doru Costache, *Humankind and the Cosmos: Early Christian Representations* (Leiden and Boston: Brill, 2021).

3   According to Greimas' germinal work *Du sens* (Paris: Seuil, 1970), the task of semiotics lies in putting sense in a condition to signify. In other words, *sense* is the given as such it is not definable, whilst *signification* is the result of a transposition. In A. J. Greimas and J. Courtés, *Semiotics: A Dictionary* (*sub voce* "Sense") it is said that sense can be considered both that which enables the operations of paraphrasing or transcoding, and that which grounds human activity as intentionality. Note that it is precisely the reference to human intentionality that differentiates semiotics from the "hard" human sciences, or from philosophies that theorise an object-oriented ontology. On the latter, see T. Morton and D. Boyer, *Hyposubjects: On Becoming Human* (no place: Open Humanities Press, 2021); T. Morton, *Hyperobjects: Philosophy and Ecology after the End of the World* (Minneapolis: University of Minnesota Press, 2013). On the so called "nonhuman turn," see also R. Grusin (ed.), *The Nonhuman Turn* (Minneapolis: University of Minnesota Press, 2015) and E. Kohn, "Anthropology of Ontologies," *Annual Review of Anthropology* 44 (2015): 311–327, DOI: 10.1146/annurev-anthro-102214-014127.

patristics is the traditional source for considering the limits and the thresholds of meaning. In any form the humanities might take—at least in principle—the object of knowledge somehow overlaps with the subject of knowledge, namely, the human being. The latter is understood at once as actant subject and actant object. But what would become of meaning if the actant subject and the actant object were no longer overlapping entities? And what would happen if the overlapping subject and object of the humanities were to occur *without* the icon of one appearing in the other? Could we still refer to the humanities as humane?

What is at stake for human sciences is the question of whether humanism is still possible. Two or three possible outcomes can be discerned: first, a *semiotic* outcome, by virtue of which the subject/object binary remains the precondition for analysing *sense and signification*, and, second, a *philosophical* outcome, which gives up the subject/object dyad in a couple of ways. In the latter case, two possibilities are foreseeable: the *internalist* approach, where the object is considered a logical and linguistic posit able to combine stimulating aspects similar in behaviour; and the *impersonalist* approach, according to which the subject is not the precondition but the product of social methods of individuation.

Come what may, the subject/object dyad is destined to condition the philosophical debate for a long time. In the following pages, the two terms will be treated insofar as they form the premise of this discussion. More interesting to me is the space where it may be possible to reach some clarity about these terms. Is it a logical space? Or is it an anthropological space? And is that space empty or not? Without presuming to explore the issue exhaustively, I will attempt to outline this "between" or betwixt space. I will show that this space is not empty, but inhabited by strange entities, that is, on the one side, the transitional objects of psychoanalysis and, on the other, the sacred icons of the Christian tradition. My intention is to propose that there is a kind of kinship between these two types of entities. An enquiry of this kind will help us grasp the place that the humanities could occupy in the near future, which many people already depict as post-humanist. Not being a hard scientist, I am nevertheless aware of the problems the subject/object binary cause beyond the humanities. Perhaps the ensuing discussion will provide answers, albeit indirectly, to issues at stake for broader enquiry, including for the faith and science interactions.

The Disappearance and the Rescue of the Subject

Knowledge requires postulating at least an object. If the theorists of object-oriented ontology were right, it could be assumed that it is not always necessary that a given subject be present. But even without going to such extremes, the presence—or the mere supposition—of an object is the necessary condition for knowledge: knowing always involves knowing *something*. That is one reason why the ancient Greek philosophers did not conceive of subjectivity the way postmodern culture does. For the Greek philosophers, the subject, ὑποκείμενος, was everything that could be spoken of or, better yet, the subject of the proposition. Nevertheless, they also called ὑποκείμενος everything that one could observe behind things; everything one would call part of the world or part of the *kosmos*; in a word, every object.[4] Whether subject or object, things in the ancient world were considered not inert but rather powerfully pulsating bodies animated by an author, artist, or demiurge.[5] Things and artefacts were capable of speech: they had a voice and behaved like emanations of their creator.[6] Echoes of this perception are clearly audible in scriptural psalms (see especially, Psalm 44). There is something poetic and magical about this intersection of animate and inanimate beings, a familiarity that the modern

---

4	The relationship between the modern concept of subject and the Greek ὑποκείμενος is critical. A correct approach to this topic features in Martin Heidegger's *Logik als die Frage nach dem Wesen der Sprache*, a work recently discovered and published as volume 38A of his *Gesamtausgabe* (2020).
5	See Marcello La Matina, *L'accadere del suono: Musica, significante e forme di vita* (Milano: Mimesis, 2017). See also Marcello La Matina, "As for God so for Sound," in *Polis, Ontology, Ecclesial Event: Engaging with Christos Yannaras' Thought*, ed. Sotiris Mitralexis (Cambridge: Clarke, 2018), 133–150.
6	Known as "Pygmalion's power." See Ernst H. Gombrich, *Art and Illusion: A Study in the Psychology of Pictorial Representation* (London: Phaidon, 1959).

world has dismissed. Indeed, beginning with Descartes,[7] the subject, as we moderns are used to thinking of it, takes revenge on things, on what, henceforth, would constitute pure extension (*res extensa*), matter or physical environment. Things, beings, and artefacts are now voiceless. Descartes' subject (*res cogitans*) becomes the *ego cogito*—a thinking subject, pure cognitive function, mind, or other impalpable reality.

The focus on the cogitating *ego* gave rise to a limitless, invisible dimension opposed to the objective external world, that is, the mind, consciousness, the computational faculty that enables the human person to build a world and to accumulate experiences, treasures of the intellect that inhabit the palace of memory. Taking its cue from this modern mindset, the twentieth century has deeply altered the meaning and forms of knowledge. New objects of study were established, more sophisticated methods of examination devised. Although these changes have impacted all branches of learning, their effect on the human and the social sciences proved to be decisive. So much so that, for about seventy years, a new scientific paradigm—one that considers phenomena as structures in a system and treats them as though they can be known as objective facts—has supplemented to the point of supplanting the traditional humanities.

As in the past, the first signs of change were seen in disciplines concerned with language and communication. To give just one example, in the 1940s Louis T. Hjelmslev envisioned a new gnoseological paradigm. In his words, "A linguistic theory which searches for the specific structure of language through an exclusively formal system of premises must seek *constancy*, which is not anchored in some 'reality' outside language."[8] Constancy, Hjelmslev argued, would have ensured the epistemological autonomy of linguistics, making it a model for other sciences. He predicted that traditional philologists and linguists would resist this new approach to language modelled on *iuxta sua principia*:

---

7   See René Descartes, *Discours de la méthode* (Leiden: Maire, 1637), and especially his *Philosophicae Meditationes*.

8   Louis T. Hjelmslev, *Prolegomena to a Theory of Language*, trans. Francis J. Whitfield (Madison: University of Wisconsin, 1961); orig. ed. *Omkring sprogteoriens grundlæggelse* (Copenhagen: Bianco Lunos Bogtrykkeri, 1943).

The search for such an aggregating and integrating constancy is sure to be opposed by a certain humanistic tradition which, in various dress, has until now predominated in linguistic science. In its typical form this humanistic tradition denies *a priori* the existence of the constancy and the legitimacy of seeking it. *According to this view, human, opposed to natural, phenomena are non-recurrent and for that very reason cannot, like natural phenomena, be subjected to an exact and generalising treatment.*[9]

The rift between the categories of subject and object has thus arrived. For centuries, humanities scholars had employed historical-critical methods of a largely circumstantial nature.[10] In this traditional view, knowing was the standard of every human deed. And it was the human being who, where knowledge was concerned, proved to be the measure—the μέτρον—of all knowledge, and of every other deliberate undertaking. Humanistic knowledge was therefore a form of human praxis (πρᾶξις). With the advent of the new human sciences,[11] the subject of conventional *studia humana* had to surrender its role as knowing agent to the objective protocols of a system. In other words, the personal *iudicium* of the philologist, or any other humanities academic, was replaced by the impersonal analysis of the new structuralist disciplines. In my opinion, constancy spelled the breaking point. Built on methodological criteria, constancy introduced the idea of repeatability into the study of human phenomena. By admitting that constancy applies not only to natural phenomena, but to human matters as well, human phenomena were implicitly stripped of uniqueness and unrepeatability.

On the subject of human judgement—and the humanistic *iudicium*—Hannah Arendt took a stand against those who argued that people had become incapable of establishing original criteria to make judg-

---

9   Hjelmslev, *Prolegomena*, 8 (italics mine).
10   See the works of Carl Ginzburg, *Spie: Radici del paradigma indiziario*, and *Miti, emblemi, spie: Morfologia e storia* (Torino: Einaudi, 1986).
11   In the 1960s, Roland Barthes identified a quadrivium of experimental sciences in the paradigms of linguistics, psychology, sociology, and anthropology.

ments, and that the best one could do was apply rules of behaviour.¹² It is worth quoting the following passage, written around the same time as Hjelmslev's *Prolegomena* and Schrödinger's *Shearman Lectures*, a few years before 1950:

> [I]f human thinking were of such a nature that it could judge only if it had cut-and-dried standards in hand, then indeed it would be correct to say, as seems to be generally assumed, that in the crisis of the modern world it is not so much the world as it is the human being itself that has become unhinged. This assumption prevails throughout the mills of academia nowadays, and is most clearly evident in the fact that *historical disciplines dealing with the history of the world and of what happens in it were dissolved first into the social sciences and then into psychology*. This is an unmistakable indication that the study of a historically formed world in its assumed chronological layers has been abandoned in favor of the study, first, of societal and, second, of individual modes of behavior. *Modes of behavior can never be the object of systematic research*, or they can be only if one excludes the human being as an active agent, the author of demonstrable events in the world, and demotes it to a creature who merely behaves differently in different situations, on whom one can conduct experiments, and who, one may even hope, can ultimately be brought under control.¹³

In no time, this new paradigm sparked reactions both for and against. And those against did not always come from the camps you would expect. For instance, people who held the structuralist revolution hos-

---

12   On Hannah Arendt's distinction of agency and behaviour and on the philosophical consequences of the prevalence of behaviour in philosophy (with reference to Greek fathers too), see M. La Matina, "Acting and Behaving: The Philosopher in Ancient Greece and Late Modernity," *JoLMA: The Journal for the Philosophy of Language, Mind and the Arts* 3:1 (2022): 7–28, http://doi.org/10.30687/Jolma/2723-9640/2022/01/001.
13   Hannah Arendt, *The Promise of Politics*, ed. Jerome Kohn (New York: Schocken Books, 1993), 104–105 (italics mine). I have compared this quotation with its version in a manuscript source held at the Library of Congress, Digital Collections, marked *Hannah Arendt Papers—Box 79—Speeches and Writings File, 1923–1975*; Essays and lectures; "Die Vorurteile," undated, sheets 022868 (–5) and 022869 (–6).

tage were not just rearguard philologists, as Hjelmselv had predicted (in Italy this group was actually enthusiastic about the methods of literary semiology),[14] but rather a large constituency of the philosophical world then engaged in debating postulates introduced by quantum physics and the theory of general relativity. Even some physicists advanced caveats of a philological and philosophical nature that could be traced back to the Greek conception of scientific thought. To take just one example, in several essays, Erwin Schrödinger—one of the fathers of quantum theory—pointed out the Greek foundations of the scientific concept of the world, in particular the postulate that the world is intelligible, and the postulate that the ability to build a scientific image of the world demands to exclude the knowing subject from the representation of the known object.[15]

A large number of philosophers also came out vehemently against the method of this new physics. In her essay *Sur la science*, Simone Weil even denounced the disappearance of modern science (*nous avons perdu la science sans nous en apercevoir*). A practice that bore the same name yet presented radically different characteristics was, she argued, surreptitiously introduced in its place (*Ce que nous possédons sous ce nom est autre chose, radicalement autre chose, et nous ne savons pas quoi. Personne peut-être ne sait quoi*).[16] What Weil sensed in the changing paradigm was the weakening of a relationship between the action of the subject and the behaviour of the studied object. She claimed that, far from expanding its cognitive practices, in the twentieth century classical science had lost something essential for doing science: "the analogy between the laws of nature and the conditions of

---

14 See Marcello La Matina, *Il testo antico: Per una semiotica come filologia integrata* (Palermo: L'Epos, 1994).
15 Cf. Erwin Schrödinger, "Quelques remarques au sujet des bases de la connaissance scientifique," *Scientia* 57 (1935): 181; idem, "Nature and the Greeks," held as The Shearman Lectures, University College, London, May 1948; now in id., *Nature and the Greeks* (Cambridge, MA: Cambridge University Press, 1948).
16 Simone Weil, *Sur la science* (Paris: Gallimard, 1946); online edition. Translation mine.

work,[17] that is, the principle itself; and it is the hypothesis of *Quantum that beheaded it*" (*l'analogie entre les lois de la nature et les conditions du travail, c'est-à-dire le principe même; c'est l'hypothèse des* quanta *qui l'a ainsi décapitée*).[18] For a philosopher as steeped in ancient Greek studies as Simone Weil, it must have been intolerable to think of κόσμος being dissociated from all the processes of ποίησις or removed from the political dimension of πρᾶξις. In truth, such a limitation was as intolerable to Weil and to Arendt as the fact that, in a world conceived of as a mechanism with no attachment to personhood, human actions could no longer aspire to be a λειτουργία,[19] a form of agency performed for the community.

The scientific and philosophical vision operative in human studies was tacitly based on an interpretation of the classical definition *homo est animal rationale*. The interpretation in question gave rise to both singularist prejudices and speciest prejudices. Singularist prejudices favour only statements concerning the individual; to use an anal-

17 As Ludwig Wittgenstein has repeatedly observed—especially in his *Philosophical Investigations*—the logical conception of language dispenses with history and consigns the definition of language to the realm of forms. Following the Austrian philosopher, I too take a stand against the Platonism of the logicians. Furthermore, I note that the topic of the relationship between language and historicity becomes particularly interesting when studying musical language. See, for example, M. La Matina, "I linguaggi e il tempo: Considerazioni filosofiche sulla storicità della Musica," *Spectrum: Journal of Music Analysis and Pedagogy* 17 (2007): 4–18.
18 Simone Weil, *Sur la science*. Translation mine. Similar statements against quantum physics can be found in Hannah Arendt, *The Human Condition* (Chicago and London: University of Chicago Press, 1958), 3: "the first boomerang effects of science's great triumphs have become obvious in the crisis of the natural sciences themselves. The trouble concerns the fact that the 'truths' of the modern scientific worldview, though they can be demonstrated in mathematical formulas and proved experimentally, will no longer lend themselves to normal expression in speech and thought.
19 I use the word λειτουργία in a very broad sense, one that is not limited to the liturgies of historical religions, although it originates from them, particularly Christian liturgies. By this word I refer to all the devices by means of which a community (or a qualified member of it) controls the conditions of truth of the utterances or actions on which its form of life depends. I have written about this—with reference to the difference between the Christian West and East—in *L'accadere del suono*, 49–63. More recently, I have returned to similar themes in the volume *Archäologie des Signifikanten: Musik und Philosophie im Gespräch* (Würzburg: Königshausen und Neumann, 2020).

ogy, it is as though, having described one flower, human and social sciences could ignore the bunch.[20] In turn, speciest prejudices consider relevant to human sciences only our species, not the diverse community of people. In a compelling passage in her *Denktagebücher*, Arendt condemned both forms of prejudice. In her words:

> The error of philosophers has always been that they thought that *Human being* relates to *people* as *Being* relates to existing *beings*; namely, the way Being, as the grounding principle, makes each existing being into a certain being; by the same principle, *human being* (namely, "Human" as an ideal type) makes existing *human beings* into certain *people*.[21]

According to Arendt, the speciest vs singularist viewpoints would arrest the development of knowledge, impeding us from grasping its authentically plural character, specifically, its political, anthropological, and ecological character:

> Because *Human being* has been used as *the Being*, the concept of *Human being* remained stuck in the representation of an animal species; ... This "ideality" derives solely from the fact that we do not yet have a concept of the human being that does not refer to animal life.[22]

If Arendt is right, the recent form of human sciences is founded on a concept of the human being that automatically assumes the speciest concept of living. Human beings are reduced to nature and behaviour, and what they look like or the sound of their voice is no indication that they might be historically significant and redeemed. But we shall return to this human being "with no image or face" (ἀνεικόνιστον καὶ ἀπρόσωπον, as the Greek fathers would have it) towards the end of this

---

20 See Byeong-uk Yi, "The Logic and Meaning of Plurals Part I," *Journal of Philosophical Logic* 34 (2005): 459–506.
21 Hannah Arendt, *Denktagebuch 1950–1973*, vol. 1, ed. Ursula Ludz and Ingeborg Nordmann (München: Piper, 2020), 128. Translation mine.
22 Arendt, *Denktagebuch 1950–1973*, 1:128. Translation mine.

study. For now, let's just point out that, almost a century ago, these premises generated a debate over the function of human sciences: even if we continue to call our studies humanities, the present epistemic situation is far different from the classical *studia humanitatis*. As often happens with such sweeping subjects, the debate developed along parallel lines—in the sciences and at the level of beliefs and opinions about our social narrative, both in our ordinary lives and in the virtual realm of social media—and, with increasing urgency, it has gripped the religious sphere.[23]

Common Sense and Philosophy on the Subject/Object Divide

Cultural movements of the second half of the twentieth century bear traces of both the old and the new paradigm. In the early decades of the twenty-first century, too, human sciences continued to link the two notions inextricably. Moreover, in many cases the original subject/object binary has been overlain with a parallel dyad, one ethically endowed: the person/thing dyad. Thus, in many discussions a kind of scientific shorthand has emerged that equates the subject with the person and the object with the thing. To say that we need to look after people more than things, and that subjects count more than objects,

---

23   This debate is ongoing. An intriguing collection of essays is *Religious Education in a Mediatized World*, ed. Ilona Nord and Hanna Zipernovszky (Stuttgart: Kohlhammer, 2017).

is commonplace.²⁴ Saying so conceals at least two truths. On the one hand, there is a kind of naïve personalism that is always unaware of its origins and aims.²⁵ On the other, there is an equally unaware curiosity about things, which reveals an unconscious idolatry of the object.²⁶ This morbid fascination with the object is often confused with virtuous λατρεία, worship—about which we shall say more at the end of this paper. The contemporary world overflows with objects, gadgets, and goods, so much so that, in order to be considered important, people

---

24   The classical reference is to the categorical imperative formulated by Kant: "So act as to treat humanity, whether in your own person or in another, always as an end and never as only a means." See Immanuel Kant, *Grundlegung zur Metaphysik der Sitten*, 1785; *Groundwork of the Metaphysic of Morals*, trans. James W. Ellington, 3rd ed. (London: Hackett, 1993), 36. On the question of being a person, see Simone Weil, *La personne et le sacré* (Paris: Rivages, 2017). The standard logical viewpoint seems to consider irrelevant the ontology of the person/thing divide; in both cases, logicians instead talk about individuals. For more on this topic, see Peter F. Strawson, *Individuals: An Essay in Descriptive Metaphysics* (London: Methuen & Co., 1959). In general terms, an object is anything that can be possessed or dismissed by subjects provided with intentionality. On the difference between having and being, the following are two classical works: Erich Fromm, *To Have or to Be?* (New York: Harper & Row, 1976) and Martin Buber, *I and Thou*, trans. Walter Kaufmann (New York: Scribner, 1937; Germ. ed. 1923). The subject/object dyad can also be seen as a relation among bodies in Foucault's sense. See Roberto Esposito, *Le persone e le cose* (Torino: Einaudi, 2014). For a consistent attempt to draw a line between objects as things and subjects as persons, see Robert Spaemann, *Personen, Versuche über den Unterschied zwischen "etwas" und "jemand"* (Stuttgart: Klett-Costa, 1996). In the end, the notion of subject/person is a timeless subject in classical and contemporary Greek philosophy, and in Christian theological debates. For example, see Christos Yannaras, *Person and Eros*, trans. Norman Russell (Brookline, MA: Holy Cross Orthodox Press, 2007).

25   In addition to naïve personalism, there also exists philosophical personalism. It was subjected to profound analyses by French existentialists. On the notion and the movement of personalism, see Emmanuel Mounier, *Écrits sur le personnalisme*, Points Essais (Paris: Seuil, 2000).

26   The idea of a "society of objects" has become widespread of late. A semiotic account of the objective/objectal topic is provided in the monographic issue of *Protée* titled *La société des objets: Problèmes d'interobjectivité* (ed. G. Marrone et E. Landowski) 29:1 (2001). Objects as consumer goods signal the ethos of contemporary society; see Emanuele Coccia, *Le bien dans les choses* (Paris: Rivages, 2013). See also Byung-Chul Han, *Die Austreibung des Anderen* (Frankfurt/Main: Fischer, 2016). The transformation of subjects into objects and of persons into things is one of the most debated topics in sociology and philosophy today. See *Tiqqun: Premiers matériaux pour une théorie de la Jeune-fille* (Paris: Mille et une nuits, 1999).

themselves frequently take on the appearance of objects.²⁷ Moreover, if people fall ill and die, things seem to remain radically indifferent to death and illness. The world is well-grounded, so goes one argument, because things provide it with a lasting life as well as objective consistency.²⁸ Hence the admonition "love people more than things," for in popular opinion people are affected by illness and therefore need more care; consequently, a world of individuals is considered not well-grounded.

Anyway, if people are subjects and things objects—and this distinction matters—then the object is what is important, what remains, what is publicly observable.²⁹ The subject, in turn, is consigned to the private sphere, to the transience of experience, or to that which appears to have no scientific relevance.³⁰ Contemporary society, governed by science, demands "objectified" thinking, an image of the world based on durable objects; such an object-oriented ontology seems to leave individual experience and subjectivity on the margins. Objects are du-

---

27 The world of electronic and digital media is often described as a world of illusions. This claim is not without foundation, especially taking into account the prophetic volume by Marshall McLuhan, *Understanding Media: The Extensions of Man* (New York: Signet Books, 1964).
28 Hidden behind the notion of object is the Greek heritage, for in the ancient Greek world the artist's and the artisan's process of production was considered analogous to the creative process of composing poems (ποιεῖν). Interesting remarks on this subject are made by Hannah Arendt, *The Human Condition* (Chicago and London: University of Chicago Press, 1958).
29 The notion of object (lat. *objectum*) is commonly referred to by the Greek word ἀντικείμενος ("what is opposite to something," "what is before us," and by extension "what-is-against"). Cf. the German *Gegenstand*.
30 According to popular opinion—one shared by most philosophers—the conventional idea of subject and subjectivity should largely depend on the notion of *homo interior*, formulated by Saint Augustine in his dialogue *De magistro*. The Augustinian concept is also associated with Saint Paul's notion of ὁ ἔσω ἄνθρωπος (2 Corinthians 4:16). For an interesting article on this topic, see Rastislav Nemec, "Some Views on 'Homo Interior' in Selected Writings of Augustine of Hippo," *Filozofia* 72:3 (2017): 181–191. The article explores the origins of Plato's ideas about the human nature up to the Alexandrian authors Philo and Origen, as well as the Cappadocian Fathers.

rable things; their birth and death are linked to their use.[31] Things are thought of as tools and not as sensitive bodies. Their duration—their lifespan, so to speak—is also measured differently. People have a date of birth and a date of death. Ordinary things do not: they "live" for as long as they are used; there is no record of them at the archives. Before cybernetics gave us smartphones and personal computers, objects did not rely much on people—or perhaps we should say that people did not rely much on objects.[32] Many believe that these new objects, which have already become an indispensable appendage of the modern subject, will influence the way people of the future, the human beings of the so called *infosphere*,[33] will be thought of and, perhaps, built.

Thus, as a consequence of the recent aforementioned paradigm shifts, the subject and the object have become in modern times the termini of human knowledge. Ever since, knowing has meant giving objectivity to the dimension of things, accounting for their stability *as* things. For instance, the debate between, on the one hand, Wilfrid Sellars and John McDowell apropos the so-called "myth of the given," and, on the other, the discussions pitting Donald Davidson against Willard Quine in regards to "inscrutability of reference" and the "third Dogma of Positivism," take place at this particular juncture.[34] If we circle back

---

[31] The notion of use (χρῆσις) was crucial to Greek philosophy, having both a political and a moral significance. The concept refers to the usage of the world as well as to the relationship between bodies or between people and texts. Echoes of the concept can be detected up to modern metaphysics. See, for instance, Martin Heidegger's discussion of *Zuhandenheit* and *Vorhandenheit* in paragraphs § 41 and 42 of his germinal work *Sein und Zeit* (Tübingen: Niemayer, 1927).

[32] In a society ruled by systems, users are the *servo-mechanism* of their own media, because these media are extensions of their body or faculties. According to McLuhan, the very appearance of this new medium could cause a sort of "numbness" similar to that of Narcissus in Ovid's myth. Such notions were introduced by McLuhan, cited above. For a discussion about the "poverty of gaze" generated by this medial numbness, see Byung-Chul Han, *Im Schwarm: Ansichten des Digitalen*, (Berlin: Matthes & Seitz, 2013).

[33] The notion of *infosphere* was introduced by Kenneth Boulding and developed by the Italian scholar Luciano Floridi in his *The Logic of Information: A Theory of Philosophy as Conceptual Design* (Oxford University Press, 2019), https://doi.org/10.1093/oso/9780198833635.001.0001.

[34] John McDowell's *Mind and World* (Cambridge and London: Harvard University Press, 1994) is required reading on this topic.

to the subject of knowledge (the *ego*, the knowing subject), we must admit that, despite the repeated and alarmed proclamations of religions and philosophies, knowledge is often produced in the same way, reduced to being a thing among other things.

In this sense, contemporary social and human sciences are nothing more than the unfolding of a drama that, from the end of the Middle Ages on, has progressively transformed the knowing subject into a known thing, to consecrate it.[35] According to Giorgio Colli, we can observe an inversion of epistemologies in recent decades: while ancient Greek epistemology dealt with the problems of knowledge in terms of objects, many contemporary epistemologies simultaneously destroy the myth of objectivity and the myth of subjectivity.[36] However, as we shall henceforward argue, there are still many practices and forms of knowledge in which a more original vision of things and their connections to people appear to be preserved. One of these forms of knowledge—as we will demonstrate later on—is psychoanalysis.

Espace Subtil: From Dichotomy to the Emergence of a Third Space

Were we to borrow an image from geometry, we might say that until now we have treated the subject and the object as the endpoints of a segment, between which we placed the line of knowledge. But let's consider of what the segment—the interval that both separates and

---

35 The debate continues. A more comprehensive depiction of the problem was drawn decades ago by Edmund Husserl in *Die Krisis der Europäischen Wissenschaften und die Transzendentale Phänomenologie: eine Einleitung in die Phänomenologische Philosophie* (Hamburg: Meiner, 2012; ed. orig. 1956). For the recent debate, see Alberto Asor Rosa, Ernesto Galli della Loggia, and Roberto Esposito, "Un appello per le scienze umane," *Il Mulino* 6 (2013); the online edition of this paper can be found at https://www.rivistailmulino.it/a/un-appello-per-le-scienze-umane.

36 Assuming that knowing is the act by which a subject constructs the representation of a given thing, Giorgio Colli is right to argue that "the Object is neither a formal nor substantial element by which one can arrive at a representation ... but rather something whose significance or reality can be clarified only if the representation is presumed." *Filosofia dell'espressione* (Milano: Adelphi, 1969), 7.

connects the endpoints—is made. We will propose three hypotheses: one logical-linguistic, one psychoanalytic, and one semiotic-anthropological. The first two preserve the subject/object dyad and will therefore be treated together; the third, however, does not, so we can develop it independently, in a way never before proposed.

Scholars know well the logical-linguistic interpretation, the subject/object couple representing a binary opposition familiar to contemporary linguistics.[37] As such, rather than a relationship, it expresses a dichotomy between the two constituent terms, with the result that knowledge is a state of the system and not a gradable process. Furthermore, once the subject and the object are counterposed as a structure in a system, they exhibit a different set of traits, more than a relationship (subject/object). We are left to decide whether it is a privative opposition, where one of the terms—bearing the distinctive mark—is called the "marked term" (marked/–marked). Certain semiotic positions recommend such a reading. According to some schools of thought, the subject is the term that bears the mark /*intentionality*/, while the other term does not. Similarly, if we consider the object the marked term, we see it as the bearer of the mark /*value*/: the object is the site invested with values in a given culture.

A common feature of this type is the negative formulation. In an oppositionist couple—just as Saussure teaches—the feature that distinguishes is also the feature that individuates. In other words, in systems of this kind, it is impossible to distinguish differentiation from individuation. The system is one of pure differences, pure negativity. If we were then to apply this structural vision to the subject/object dyad, the logical space around the dyad would be (non)gradable (there would be no intermediary trait between the subject and the object) and therefore we would have no observable phenomenon to place between the constituent terms, which are mere fictions of binary logic. At most, the opposition in any context could be neutralised; this would result in

---

37 See Ferdinand de Saussure, *Cours de Linguistique générale*, 5th edn, ed. Ch. Bally, A. Sechehaye, and A. Riedlinger (Paris: Payot, 1915). For a critical consideration of binaries, see Roland Barthes, *Eléments de sémiologie* (Paris: Seuil, 1964), III.3.

the disappearance of difference between subject and object, determining a state of exception. This may be the most compelling hypothesis, since it calls to mind the political utopias aimed at installing regimes of knowledge completely devoid of differentiation—and therefore of individual actants. Thus, the logical-linguistic interpretation does not help us understand the role of the subject/object couple in the field of human science that we intend to examine and reformulate.

Psychoanalytical research into intermediary entities between the subject and the object appears to be more promising. In fact, it demonstrates that something lies between the subject and the object that is neither an object nor a subject. In order to talk about this third element, we have to introduce a new concept, the so-called transitional object. What is a transitional object? Before we proceed, we must first identify a few psychoanalytical terms. Let's begin by pointing out that scholars agree that there is a difference between ontogeny and phylogeny; that difference matters here. It is often said that an individual relives the history of all humanity in its own development. But, of course, that statement is not always true. In fact, it is more proper to talk about history in connection to individuals and their existence. Species have no history, properly speaking. If we accept this distinction, then talking about a "history of objects" is not the same in human individuals as it is in the human species. Taking this distinction as a given, let's turn our attention to the ontogenetic side of the subject/object couple.

What specifically happens during human development? On its ontogenetic path, a human individual—every human subject—is born without objects. As modern psychology states, a child is born as a rational subject only within the "subtle space" (*espace subtil*) where the dominant presence of the mother nullifies the need to seek objects. If anything, objects are occasionally convoked as forms of offsetting. Said differently, a child enters a space of objects only in cases where it experiences a lack of personal presence. Later on in an individual's life, the object becomes an intermediary zone between the subjectivity of the person and the objectivity of the thing. In 1972, Jacques Lacan introduced the so-called "Objet (*a*)" into psychoanalytic theory, a distant echo of Freud's drive object and Melanie Klein's partial Object. Something becomes an "Objet (a)" only when it is in a seeking relationship (quête), for it expresses the *objet-cause* of desire.[38] This, Lacan explains, displays traits that we shall call semiotic, because they can be traced back to the moment of enunciation: "Insofar as it is selected in the appendages of the body as an index of desire, it [namely, the 'Objet (a)'] is already the exponent of a function, of the index pointing to an absence."[39]

Lacan's contribution aside, the history of transitional objects stems from other scholars.[40] The name and concept first appeared in Donald Woods Winnicott, but the version of the transitional object as we will be referring to it hereafter belongs to Françoise Dolto. Winnicott's transitional objects pose a challenge to the traditional ontology that separates people from things and gives beings a different status. Its discovery, and the innumerable ways in which this notion can be applied, persuade us that within the transitional object lurks a portion of history that we might call a "wordless mythology": a universe where, instead of words, the tale is formed of images; a tale whose heroes are things, or objects, and in which subjectivity is absorbed by the thing and, in a sense, objectified without being transformed into an object. On the other hand, the transitional objects that Dolto and then Denis Vasse have in mind are not actual objects; they can embody an object, but they remain floating signifiers. They are neither denoted objects

38 Jacques Lacan, *Autres Écrits* (Paris: Seuil, 2001), 379. Lacan discusses this topic in *Séminaire XV* (1967–1968, unpublished). I have consulted a summary of this seminar.

39 Jacques Lacan, "Remarque sur le rapport de Daniel Lagache," in *Écrits* (Paris: Seuil, 1966), 647–684, esp. 682. I would like to add one comment about "Object (a)." Because it is a purely linguistic creation, this unrepresentable object seems similar to ξ, the character used by Gottlob Frege: ξ is not exactly part of symbolic language, but instead the index of the temporal staging in the analytical step-by-step construction of a sentence. In short, a kind of transitional object. Michael Dummett describes the symbol ξ as "merely a device for indicating where the argument-place of a predicate occurs." See his *Frege: The Philosophy of Language* (New York: Harper & Row, 1973), 16.

40 The bibliography of transitional objects is endless. See Denys Ribas, "L'œuvre," in *Donald Woods Winnicott*, ed. D. Ribas (Paris: Presses Universitaires de France, 2003), 35–109; Victor Smirnoff, "La relation d'objet et le vécu infantile," *La psychanalyse de l'enfant*, ed. V. Smirnoff (Paris: Presses Universitaires de France, 1992), 183–292.

nor denoting signs; they are signifiers, tasked with conjuring up the presence of a person (and not the presence of an object—be it a *Bedeutung* or meaningful babble).

How can we represent the developmental condition of floating signifiers? Infants, even absent their mothers, are never alone. Infants live in relation, σχέσις,[41] as the Greek fathers would call it. Let us define σχέσις as the relation that precedes and forms the phylogenetic foundation of every successive appearance of the subject and object. To support this interpretation, suffice it to quote Dolto's assertion that infants "invent this relation" and "conjure the presence of their mother by babbling, convinced that they are repeating the phonemes that they had heard their mother utter and, thus persuaded by this trick, feel not alone but rather for and with her (*pour et avec elle*)."[42]

An interesting feature of σχέσις is the condition of indiscernibility between mother and infant, a condition Dolto calls *mêmeté d'être* (ontological memory, memory of being, or even sameness of being). In this condition, the child tries to stay in contact with the primary object—mother—by producing expressive (hence not objectual) simulacra. The child babbles, gesticulates, expresses itself in several ways. The child appears within, *not in*, a relation with the object. The proof is that its utterances neither refer to a *Bedeutung* nor can they be interpreted as acts of reference. Rather, they are signifiers in relation with other signifiers. They are signifiers that we would like to call *echo-like*, reformulations (transpositions, recreations) of enunciation acts that

---

41 When I refer to the ancient Greek word σχέσις, rather than the classical Aristotelian notion, I have in mind the idea of "relation" in the Greek patristics of late antiquity, for ex. the Cappadocian Fathers. For more on this subject, see Ilaria Vigorelli, *La relazione: Dio e l'uomo: Schesis e antropologia trinitaria in Gregorio di Nissa* (Roma: Città Nuova, 2020). See also Marcello La Matina, "God Is Not the Name of God: Some Remarks on Language and Philosophy in Gregory's *Opera Dogmatica Minora*," in *Gregory of Nyssa: The Minor Treatises on Trinitarian Theology and Apollinarism*, ed. Volker H. Drecoll and Margitta Berghaus (Leiden and Boston: Brill, 2011), 315–336.

42 Dolto, *L'image*, 35. It is the mother who, with her words, mediates the absence of an object for the benefit of her infant; in technical terms, as Lacan would have concurred, the partial object is evoked by the total object (Dolto, *L'image*, 64).

can be assigned to a different space every time. They are not just evocations—they are affirmations of the (imagined) presence of the mother.

If this interpretation is correct, it is easy to think that the transitional objects discovered by Winnicott introduce what I shall call a proximal zone. Infants choose these non-objects from their immediate surroundings and therefore the non-objects are enabled to establish a transition between the original relationship with the maternal breast and the constitution of real objects in the external world.[43] A similar view is taken in Dolto's discovery of transitional objects. These objects can suggest something interesting about the human subject/object divide, so that many popular, deeply ingrained beliefs must be rewritten. In particular, thanks to psychoanalysis, before appearing as tools or products, objects function as what I call "floating signifiers";[44] they are not rooted in conceptual grammar nor do they refer to semantics structured by conventions. Rather, they have their basis in the child's bodily image; they are firmly grounded in its personal history. Notice that, in spite of its name, the image is less a visual formation than a *tensive-muscular habit*. Moreover, the transitional objects are witnesses and places where the category of mediation is applicable. In this sense, they are also the place where the desire for a relationship appears in the form of desire for the Other (*désir d'Autrui*). In this sense, it is always a relation with the Distal Other.

Transitional Field and Anthropic Zones

The third interpretation of the space between subject and object is semiotic-anthropological. This consists in rewriting some previously discussed theories of transitional objects. It is expedient to stress that transitional objects should be considered neither objects nor pseudo-objects. To me, they seem more like signifiers that have yet to be

---

43 See the innovative article by D. W. Winnicott, "Transitional Objects and Transitional Phenomena: A Study of the First Not-Me Possession," *International Journal of Psycho-Analysis* 34:2 (1953): 89–97.
44 For more on this subject, see Claude Lévi-Strauss, "Introduction à l'oeuvre de M. Mauss," in Marcel Mauss, *Sociologie et Anthropologie* (Paris: Presses Universitaires de France, 1950).

caught in the net of grammar and are therefore drifting in the *espace subtil* inhabited by mother and infant. The following is an important observation: if we connect Winnicott's analyses to Dolto's, it turns out that transitional objects reside in the region between subject and object, but do not belong to merely one zone situated between subject and object. Therefore, hereafter we can do without the labels "subject" and "object," and call this region *transitional field*.[45]

First, we must say what the transitional field is. The signifiers that operate in the transitional field do not settle into fixed patterns: sometimes they refer to proximal signifiers, sometimes to superimposable signifiers, and still other times they evoke signifiers that cannot be placed in either the superimposable sphere nor in the adjacent sphere. When this happens, they evoke a distal (an ancestor, a mythic time or space, a Freudian thing, etc.). Moreover, sometimes the reference is spatial in nature, and other times it is not. Which is why it is important to articulate the transitional field semiotically. Therefore, the transitional field is the semiotic *miniverse* that expresses proximal space and connects it to the distal space evoked.[46] It is about understanding the nature of the transitional field is and about distinguishing the phenomena associated with it.

---

45 The view proposed here is not a variant of the well-known logical Platonism. Firstly, I speak of transitional objects as signifiers—bodies—sensible things linked to the corporeity of the human person. Secondly, as I have written in my *Archäologie des Signifikanten*, these signifiers are assimilated here with Christian icons, with which they share a perspective directed not to the past but to the future. It is worth referring here to a passage by Ps.-Maximus the Confessor (*Scholia in librum De Ecclesiastica Hierarchia*, PG 4, 137A–D), where he writes that "truth is the state of things to come" (ἀλήθεια δὲ ἡ τῶν μελλόντων κατάστασις). John Zizioulas notes: "In this passage, Saint Maximus interprets in his own way the concept of Eucharist as image and symbol in relation of the concept of causality ... The divine Eucharist is for him an image of the true Eucharist which is nothing other than 'the state of things to come.' The truth of 'what is now accomplished in the *synaxis*' is to be found not in a Platonic type of ideal reality, but in a reality of the future." J. Zizioulas, *The Eucharist and the Kingdom of God*, trans. Elizabeth Theokritoff (Alhambra, CA: Sebastian Press, 2022), 21–22.

46 For a discussion about the proximal and distal emissary, Marcello La Matina, *Cronosensitività: Una teoria filosofica per lo studio dei linguaggi* (Roma: Carocci, 2004).

François Rastier notes that in every culture there exist significant disruptions to the contiguity of *Umwelt*. For example, he has shown homologous positions along the four axes (person, time, place, and mode). Different languages may have different names for these axes, yet there can only be three zones traced, which in his most recent work he describes thus: the *identity* zone, where the subject establishes coincident rules with self-image; the *proximal* zone, where the subject is adjacent to empirically accessible entities (what I call signifiers); and finally the *distal* zone, situated in another time and space that by their mode transcend the first two zones.[47] Thus, in every language, we have a first, second, and third person, just as we have past, present, and future, and other aspects to which, though they vary, speakers constantly refer. Table 1 breaks down these gaps and homologies between Rastier's three zones:

	Identity Zone	**Proximal Zone**	**Distal Zone**
1. Person	I – We	Thou – You	He/She – It – One
2. Time	Now	Once – Soon	Past – Future
3. Space	Here	There	Over there – Somewhere else
4. Mode	Certainty	Eventual	Possible – Unreal

Table 1

This theory makes an important point about the connections between zones. Rastier identifies two: the empirical *couplage* (or linkage between the identity and the proximal zones); and the transcendent *couplage* (between the first two and the distal zone). Note the terminology and subdivisions in Figure 1 (by Rastier): *couplage empirique* (empirical nexus), *couplage transcendant* (transcendental nexus), *zone identi-*

---

47   I am primarily referring to François Rastier, "Représentation ou interprétation? Une perspective herméneutique sur la médiation sémiotique," in *Penser l'esprit: Des sciences de la cognition à une philosophie de l'esprit*, ed. V. Rialle and D. Fisette (Presses Universitaires de Grenoble, 1996), 219–253. The figure is at 246.

taire (identity zone), *zone proximale* (proximal zone), *zone distale* (distal zone), *frontière empirique* (the empirical border), *frontière transcendante* (the transcendental border), *fétiches* (fetiches), and *idoles* (idols).

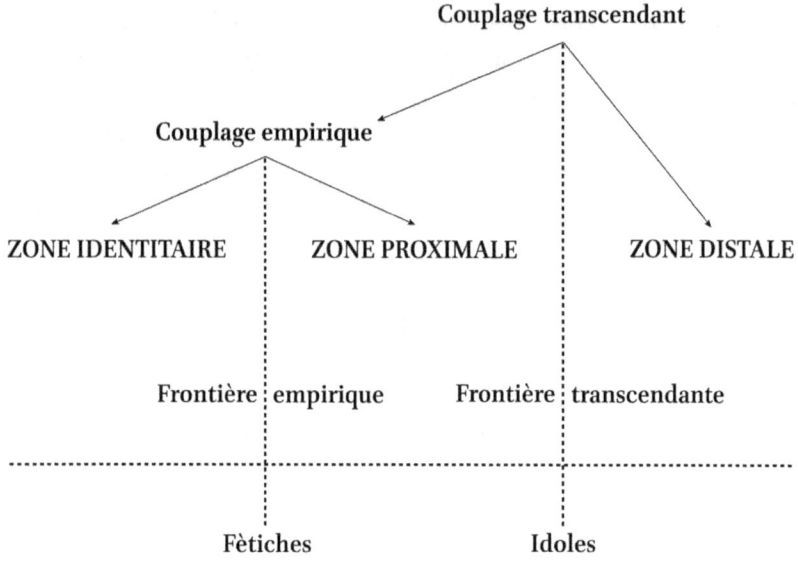

Figure 1

Rastier believes the distinction between the first two zones and the third is significant. The objects present in empirical space are called fetiches (charms) and those in transcendent space idols. If we attempt to apply this new terminology to the matter at hand, we might ask ourselves what category the Byzantine icons or their counterparts, ὁμοιώματα, fall into, based on their mode of being. The icons cannot be charms, since no "magical" power is attributed to them. Nor are they idols, since they are not worshipped like divine effigies. The icons seem to conjure a mode of being that is not included in Rastier's diagram (or, if it is, it is so incongruously). Strictly speaking, not even transitional objects seem to fit neatly into these categories. In fact, they shuttle between the empirical and the transcendent, without being able to connect these *signifiants flottants* to any regularity provided by a code or connected to some standard significance.[48] We shall soon turn to this question.

Renunciation of the Object and the Proto-Sacrifice of the Child

There are no objects in the transitional field; hence there is no space for the semantic reference (which always requires a referent or *Bedeutung*). Here, on the contrary, as Jacques Lacan beautifully puts it, every signifier "represents a subject for another signifier." Everything happens in the space of air, in the space of breath that binds mother to child, or the child and the maternal ghost, by means of speech sounds and rhythms that evoke the physiology of feeding, hunger, expulsion, and crying. Above all, everything happens by means of the first clumsy attempts to reproduce those sounds that make the mother present when she is absent, when she is far away. The space between subjects and objects is now an all-embracing, bodily space: "This way, we understand that language is not an immaterial abstraction, but rather the body of the infant perceived in the network of signifiers, its subtle body, truer than the opaque materiality of a meaningless organism. In this sense, the word of the mother (and of others) gives body to the child."[49]

---

48   One could also word the question differently: "On what basis can we decide when a Transitional Object falls into the identity space of the subject or into a proximal or distal space? Studies show that this introduces a zone adjacent to the subject, but sometimes it slips into a distal space/time. Clinical data about this have not resolved the matter. What counts for the birth of a Transitional Object is its *aspectuality* (that is, whether it has a continuity of repeated and recognised perceptions that the child can organise its bodily *imago* around). In the subtle space of signifiers, the bond between mother and child, severed together with the umbilical cord, can be reconstructed thanks to the presence of these *motherised* objects (*objets mamaïés*), i.e., things capable of conjuring for the infant a memory of the mother's reassuring presence." Dolto, *L'image*, 70. Translation mine.
49   Denis Vasse, *L'ombilic et la voix* (Paris: Seuil, 1974), 67–68. Translation mine.

In the transitional field signifiers come in all sizes. They are comparable to images (εἰκόνες); how so will be explained later on. I have set aside the problem of establishing what kind of images give form to the transitional field. Images that function as unconscious images of the body are, in any case, preponderant. Unlike the claims of some schools of psychoanalysis, it appears to me that the bodily image should not be considered a mere projection of the child: if that were true, it would be an ontic phenomenon lacking ontological depth. Instead, I believe that the bodily image (εἰκών connected to the "reticular" story of signifying bodies) should be seen as a relational—and more importantly historical—phenomenon, for which what counts is the uniqueness and unrepeatability of every *couplage* between actant and *Umwelt*.

In simpler terms, I believe that the infant is never alone, but always caught in the web of signifiers that turn the infant's originating Other into an *allelon*.[50] What I am saying appears to confirm Dolto's intuition that "the image of the body is always a potential image of communication with the phantom. Human solitude is never unaccompanied by a mnemic trace of a past contact with either an anthropomorphised other (*autrui anthropomorphisé*) or a real one."[51] If my questions are plausible, what kind of relation is the transitional field that every transitional object opens between proximal and/or distal *allelons*? And what does this new vision teach us about how to understand the epistemic and anthropological divide based on the contrast between subject and object?

All this also tells us, however, something more interesting concerning the topic: in the infant's experience, language is similar to *hierotopy*, which includes the signifier, both bodily and symbolic. No actual objects are given; only corporeal signifiers that enact a genuinely liturgical action, a sacrifice of sorts. How is such a sacrifice possible? If Vasse is right, I could be so bold as to say—and this is my thesis—that

---

50 The word *allelon* (not to be confused with "allele") is my own term. It comes from the ancient Greek reciprocal pronoun ἀλλήλων, which appears in various phrases. For a detailed explanation of the theory, see my forthcoming article, appropriately titled, "Alleli e allelouchìa: Semiotica e forme di vita."
51 Dolto, *L'image*, 35. Translation mine.

the infant's surrender of the object functions as the child's "sacrificial gesture." It is an offering made in order to compensate for the unbearable absence of the mother's body. Would it be correct to speak of a *proto-liturgy*? Vasse and Dolto would explain the behaviour of the infant by way of *mêmeté d'être*. In simpler terms, the infant blurs the absence of and desire for the maternal body. For this reason, the child literally becomes other; to compensate for the maternal absence, it would put itself, totally unconsciously, in a state of exception. Unable to cope with the mother's absence, it would seek the *mêmeté d'être*, the sameness of being with the mother, transitionally recreating the web of maternal signifiers: *Il tente d'être autre pour demeurer même* (it attempts to be another in order to remain the same).[52]

While embracing this subtle analysis, I prefer to think in semiotic terms. In the situation just described, no actual actant appears; we might second Greimas and say that we are in the presence of the *en deçà*, under the true signifier. The transitional field is the space where performing a symbolic sacrifice enables the participants to claim a form of proto-actantiality. Also indicative of this is what emerges from the studies of Vasse, a psychoanalyst who picked up where Dolto left off. Vasse considers central to this process of the mother's presence/absence what he calls the "deferred reconnection" (*rétablissement différé*) of the missing object.[53] The infant calls on the imagination, conjuring up a past experience that is felt, however, as an experience capable of launching a new future. Equally, argues Vasse, during this imaginative phase, the possibility of deferring the moment of satisfaction is introduced. This phase he calls the renunciation of the object: "At the same time that the possibility of deferring the moment of restorative satisfaction, of renouncing the object, is introduced, the subject's desire for

---

52 Vasse, *L'ombilic et la voix*, 77. Under attack is the concept of the individual, squeezed between the personal and the intersubjective realms. The topic has been the subject of many astute analyses. See the essays by P. Veyne, J. Vernant, L. Dumont, P. Ricœur, F. Dolto, F. Varela, and G. Percheron in *Sur l'individu: Contributions au colloque de Royaumont* (Paris: Seuil, 1985).
53 Vasse, *L'ombilic et la voix*, 77. Translation mine.

something other than the thing, the encounter, arises, supported by the memory traces of previous experiences."[54]

If my argument is plausible, then Vasse's renunciation of the object can be seen as a proto-liturgy or a proto-sacrifice—indeed it takes the shape of a symbolic evocation (i.e., via signifiers) that introduces the primary signifier into the proximal space: the mother (who, however, lies in the distal space). Let us delve a little longer on the renunciation of the object;[55] it raises questions about the nature and function of the transitional field. What type of space is it? Is it a logical space, as we learned from the theory of proposition? Or is it an anthropological space, tasked with mediating between identity and proximity? Or are we encountering a utopian space where we ought to place the operations that make the conjured distal signifier accessible (the mother, the lost object, the Freudian thing, etc., as well as the first signifier that goes back to God for other signifiers)? Surely—and now I can say so—it is not merely a psychic space, it is not just the fiction on which imagination and reality hinge. The renunciation of the object performed by the infant makes clear that the transitional field is homologous to the symbolic field and to the relationship between the *allelons*, which I have called *allelouchìa*; all this demands a more in-depth theoretical study, which—surprisingly—could come from semiotics more than from psychology.[56]

Before moving on to the next topic, it is worth underscoring once more the paradoxical feature that up to this point we have been making. Psychoanalysis (Lacan, Dolto, Winnicott) discovered transitional objects and shone a light on their communicative and expressive function. Nevertheless, it is in semiotic terms (and not psychoanalytical

---

54 Vasse, *L'ombilic et la voix*, 77. Translation mine.
55 Vasse, *L'ombilic et la voix*, 77.
56 My observations are not contradicted by François Rastier's consideration of the transitional object as the first model of the cultural object. In fact, the objectivity of the transitional object obeys subjective laws (*obéit à des lois subjectives*). See his "Prédication, Actance et Zones Anthropiques," in *Prédication, Assertion, Information*, ed. F. Rastier, M. Forsgren, K. Jonasson, and H. Kronning, Acta Universitatis Uppsaliensis, coll. Studia Romanica Uppsaliensia 56 (Stockholm: Almquist et Wiksell International, 1998), 443–461.

terms, as Vasse rightly observed) that one can explain the birth of the object as such, linking this emergence to the formation of a network of signifiers that occupy the subtle space (*espace subtil*). Our subject is now the transitional field, the site where signifiers of the body and language appear, the site of transformations and ritual practices and, in all frankness, liturgical practices. The transitional field is above all the space of the proto-sacrifice, the space of the signifier, where the infant renounces the object, having been forced by the absence of the mother to remake herself and experience "being alone while someone else is present." As Winnicott observed:

> Although many types of experience go to the establishment of the capacity to be alone, there is one that is basic, and without a sufficiency of it the capacity to be alone does not come about; this experience is that of being alone, as an infant and small child, in the presence of mother. Thus, the basis of the capacity to be alone is a paradox; it is the experience of *being alone while someone else is present*.[57]

Anthropic Zones and Byzantine Icons

The question of icons might not appear important to a discussion of the ways of knowledge and the redefinition of human studies, yet in fact it is.

In the philosophical vision of the Greek fathers, the subject/object dyad does not have a real theoretical purpose, whereas questions of knowledge, especially about God, are often treated by applying the idea of πρόσωπον, or, in modern terms, person. When the term πρόσωπον was first introduced, it did not have a clear semantic definition. Around its lexical, textual, and theoretical history, much literature has sprung

---

[57] D. Winnicott, *The Maturational Processes and the Facilitating Environment: Studies in the Theory of Emotional Development* (London and New York: Routledge, 1958), 29.

up, literature that we cannot give a full account of here.[58] In the texts of the Cappadocian fathers or Maximus the Confessor—those I think I know best—the term πρόσωπον comes up frequently, including as a synonym of ὑπόστασις. As I attempted to show in an earlier essay,[59] there are contexts in which it appears as a meta-indexical sign used in reference both to the context and to the co-text. In all of these cases, πρόσωπον fulfils its role when it refers to what one finds when *faced-with-a-face* and, for that reason, offers a "face" to the "face" that is looking or being looked at. In Greek patristics, the term was given a specific meaning, so that "what one is faced with" (which we might call ἀντίον) is not referred to as πρόσωπον in its ontological dimension (οὐσία) but is referred to as present-in-the-face-of-us (παρουσία). This what-is-before-us can arise either from absence, as something that did not exist before, or from ignorance, as something that emerges from oblivion or comes out of hiding. The moment ἀντίον is present, then the mode of existence (τρόπος τῆς ὑπάρξεως) of πρόσωπον is realised. To paraphrase Christos Yannaras, the πρόσωπον is nothing if not the way of ecstatic existence itself: an *existence-before-that-which-is-Other*.

Therefore, what πρόσωπον realises is a mode of existence within an anthropic zone and not within a logical-linguistic space. Whether we are talking about an icon of the Pantocrator or the icon of the Mother of God or of a saint, the mode of existence of πρόσωπον is completely different from an object trapped in the subject/object dyad. Again, Yannaras puts it well when he writes that "the person [i.e., the πρόσωπον] in its ecstatic reference—that is, in its otherness—transcends the objective properties and common signs of recognition of the form, and consequently is not defined by its nature."[60] The πρόσωπον is an ecstatic reality open to the surrounding space, the *Umwelt*. According to Byzantine anthropology, the πρόσωπον is distinct from its nature. This

---

58 For a bibliography of the idea of πρόσωπον, see Johannes Zachhuber, *Human Nature in Gregory of Nyssa: Philosophical Background and Theological Significance* (Leiden: Brill, 2000); Lucian Turcescu, *Gregory of Nyssa and the Concept of Divine Persons* (Oxford University Press, 2005).
59 La Matina, "God is not the Name of God," 315–335.
60 Yannaras, *Person and Eros*, 25–26.

double order—person and nature—is native to Greek and Oriental patristic philosophy.

In theoretical writings about icons—from Gregory of Nyssa to Maximus the Confessor and from Dionysus the Areopagite to Theodore the Studite—we often encounter the word πρόσωπον. It means, quite specifically, a relationship between the ὁμοίωμα (likeness), called εἰκών (icon, image), and the πρωτότυπος (prototype).[61] Such a relationship is placed in the space of prayer, and therefore a *hierotopic* space. The veneration of icons is not, as we know, an act of adoration, so that the icon is neither a charm nor an idol, to borrow Rastier's terms. Προσκύνησις (veneration) is, instead, a semiotic act that—to use Lacan's beautiful phrase—articulates meaning by convoking a signifier capable of signifying a subject for another signifier. In that sense, we can say that προσκύνησις validates the presence-of-person recognised as signifier via personal devotion. No wonder the second Council of Nicaea accurately drew a distinction between προσκύνησις and λατρεία (adoration, worship).[62]

In the Byzantine world, the perception of the icon is extremely close to the realm of the person, with which it often coincides. Theodore the Studite says that "icons are sometimes referred to as 'icon of such-and-such' and sometimes they are referred to as if they were the person itself, that is, the archetype."[63] The Byzantine icon is thought of neither as an aesthetic object nor a material object, but rather ὁμοίωσις (likeness), the presence of the absent one. This ὁμοίωσις renders the relationship with the archetype effective for bringing about προσκύνησις, creating a kind of *objet mamaïsé* in the *espace subtil* of liturgical devotion. One last observation: all the sources emphasise that σχέσις (relation) happens without the involvement of the object in its materiality (ἔξω τῆς ὕλης, outside of matter).

---

61 See Theodore the Studite, Ep. 57, in *Theodori Studitae Epistulae*, ed. Georgios Fatouros, Corpus Fontium Historiae Byzantinae – Series Berolinensis 31 (Berlin: de Gruyter, 1992), 164.
62 See Theodore the Studite, Ep. 57, at 167.
63 Cf. Theodore the Studite, Ep. 301, at 442. A contemporary philosophical correspondent of this idea can be found in Nelson Goodman, *Languages of Art: An Approach to a Theory of Symbols* (Indianapolis: Bobbs-Merrill, 1976).

Much can be said on this subject, but my ambitions are more modest. In fact, I shall merely suggest a possible typological kinship between the transitional field (as signifying space that ties the proximal to the distal) and the hierotopic space of the icon. Because it evokes the dimensions of time past and time future, the transitional image represented by εἰκών introduces a break in the continuity of the psychic present that encroaches on a dimension that we might call analogical or modal. This indicates that in human cultures there is a distal zone that is constructed differently from those that preceded it. Technically speaking, we should talk about the relationship (σχέσις) between a likeness (εἰκών or ὁμοίωμα) and its prototype (ἀρχέτυπον or πρωτότυπον) as a relationship between signifiers, none of which is a charm or idol. Thus reformulated, the three anthropic zones form the transitional field, which is not supported by dichotomous logic, but rather functions iconologically.

My diagram below (Figure 2) honours the work of Rastier, but revises it in part. The upper half shows the threefold division according to the arrangement of Rastier's three zones. The lower half, in turn, shows the arrangement of the transitional field as I see it. The two models are not mutually exclusive; they can be employed to describe different ontological commitments. For example, the lower half shows how the relationship between the identity zone and the proximal zone is characterised by an openness to the other and is, therefore, an ecstatic *couplage* (ἐκστατικὸν συνδυασμόν). This interpretation places the person, πρόσωπον, at the centre of the relationship with the other.

There is no subject/object relationship where "otherness refers not only to objective beings and other persons, but is also actualised principally with regard to the natural individuality of personal existence."[64] My interpretation of the relationship between the proximal and the distal zones is also considerably different. It takes the form of an analogical *couplage*, based on iconic semiosis (εἰκών recalls its prototype, πρωτότυπον). The signifiers that appear in this *couplage* are not—as I just said—idols, but icons (εἰκόνες), signifiers placed in

---

64 Yannaras, *Person and Eros*, 27.

relation with the Face, with πρόσωπον. I call this *couplage* (ἀναλογικὸν συνδυασμόν) analogical because its function reveals an aspect of reference from signifier to signifier, according to the modes of the analogy. Yannaras perfectly captures the sense of an anthropology based not on a subject/object antinomy, but on a solidarity between the person and the icon as signifiers. In his words, "if we accept the human person as the 'horizon' of the disclosure of beings ... knowledge becomes the experience of the disclosure within the context of the person's relation to objective things"; and again: "The Icon is the signifier of personal relation."[65] These two propositions encapsulate the search for a model of human studies that respects the personological and the iconological dimensions.

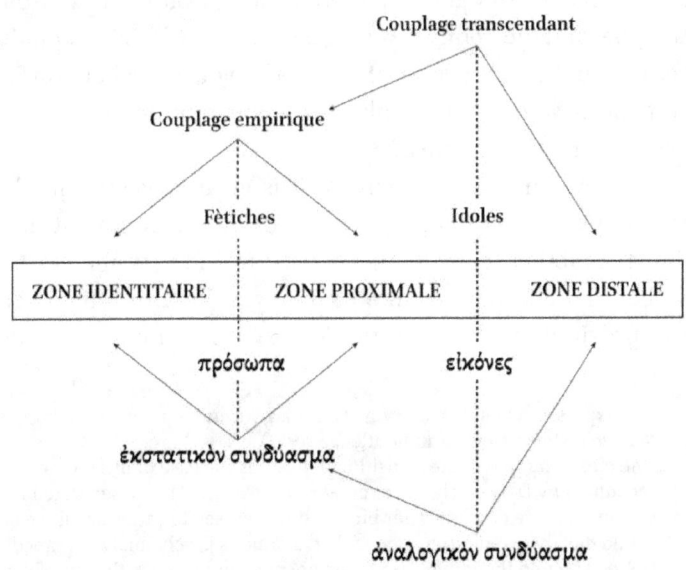

Figure 2

I would observe that all human and social sciences would greatly benefit from the application of these semiotic-anthropological categories.

---

65   Yannaras, *Person and Eros*, 184.

For example, Winnicott's point about the reassuring power of the transitional object—which constitutes "a vehicular unit," corresponding to "Linus' blanket"[66]—makes one realise that the transitional object overlaps with the newborn and is therefore an element of the anthropic identity zone. In turn, Dolto emphasises the forms of communication that take place in the *espace subtil*, characterising this space as an adjacent zone between the mother and the child. Hence the transitional object springs from the proximal zone. Dolto defines this as "an object that joins the infant to the tactile images of the foundational zone"; that is, something closely linked to the heterogenous zones and the space of communication between mother and newborn.[67] Still, in other cases the transitional object appears to be the *atmosphere* where fragments of the sensory life of child and mother float. Dolto writes: "You could say that, beyond the bodily distance between newborn and mother/wet nurse, the subtle perception of scent and voice is what continues to act, for the newborn, as the place—the surrounding space—where it observes the mother's return."[68]

At this point of the analysis, it is clear that the initial σχέσις is the *instance of enunciation*, first manifested with the cutting of the newborn's umbilical cord. With the removal of the umbilical cord, the infant body is reborn into a new economy, going from "liquid contiguity and proximity with the mother's body" to an impulsive autonomy

---

66   The expression belongs to Erving Goffman, *Relations in Public: Microstudies of the Public Order* (New York: Routledge, 1971).
67   One clear example of the transitional object is the case of little Agnes, recounted by Dolto early in her career. In 1944, after being separated from her mother just five days after her birth, Agnes refused to eat. Fearing the child would die, the paediatrician consulted a famous psychoanalyst, Françoise Dolto, who told the father, "Go to the hospital and bring with you a shirt that your wife usually wears, but make sure the shirt still bears her scent. Wrap it around the child's neck and give her a feeding bottle." Although it seemed strange at the time, Dolto's advice turned out to be sound, because the "thing" was not simply a thing, but an object capable of mediating between mother and baby. For this case, see Sophie Marinopoulos, "De l'objet « mamaïsé » de Françoise Dolto à l'« objet transitionnel » de Donald W. Winnicott," *L'école des parents* 621:6 (2016): 41–52. Cf. Dolto, *L'image*, 66–67.
68   Françoise Dolto, *L'image inconsciente du corps* (Paris: Seuil, 1984), 69. Translation mine.

made up of rhythmic events: inhalation and exhalation, nutrition and excretion, presence and absence (of the mother).[69] There is something paradoxical about this story of scientific discovery. Proceeding from psychoanalytical observations, I have arrived at the semiotic dimension and introduced the idea of the transitional field. Within that field, the rhythmic alternation of signifiers takes place (presence/absence of the mother, nutrition/excretion, satiety/hunger, etc.) and the signifying space[70] is created, where later on the subject/object dichotomy is established. At this stage of development, before the appearance of things, objects, and the *denotations* of words, all that is found in the transitional field are signifiers. It is to the web of signifiers that the newborn entrusts the work of reconstructing "the feeling of bodily fulness that necessarily connotes presence."[71] The transitional field is a web of floating signifiers and not a field of things or referents, or what linguists normally call the "signified" or the meaning. The field has the typical character of associations between signifiers. Moreover, the pre-eminence of the signifier points to a rejection of the object, or at least the deferral of "the moment of restorative satisfaction," on the part of the infant.

Icons at an Exhibition

I would now like to comment on the relationship between transitional objects and artistic language, focusing on several images from the catalogue for "Transitional Object Project Zero," the first Italian art exhibit created with the intention of collecting images of transitional objects as reproduced by artists and other creators.[72] In short, a select group of artists were asked to draw their own transitional object, as remem-

---

69   Cf. Vasse, *L'ombilic et la voix*, 67.
70   Vasse, *L'ombilic et la voix*, 69.
71   Vasse, *L'ombilic et la voix*, 76.
72   "Project Zero" was launched by Elena Cesaretti and Alessia Porfiri, designers, visual artists, and art therapists. The catalogue was edited by the artists themselves and included a preface by Marcello La Matina (Macerata: Trob, 2022). The objects mentioned in this section can be found at https://en.trob.space/gallery (accessed 12 August 2022).

bered by the interviewees; the resulting drawings and photos have become the exhibition.

I would also like to describe a connection between how transitional objects function in therapy and how icons function in prayer and veneration. What can the artists' drawings tell us about transitional objects? We know that infant psychoanalysis commonly revolves around interpreting the drawings and clay models of young patients. But in this case, we are dealing with adults whose products are not transitional objects; they are visual or sculptural sketches of them. They are not *objets mamaïsés* but personal transcriptions of a music that only the listener or the performer can know. How can they serve this discussion? Well, I am convinced that the drawings and photo collages, though not applicable to clinical study, can provide a seed for philosophical reflection. Here is my argument.

We have seen above that transitional objects are not actual objects, but signifiers floating around a transitional field, that they do not acquire specific material shape, and that their formal properties may not be defined. By virtue of their "signifying nature," it is difficult to represent transitional objects as beings, since they cannot be exhibited as normal, average-sized objects. Nor can we represent them pictorially as objects, strictly speaking. And yet, as a philosopher would say, if you can't show Being itself, you can at least attempt to show the spirit of being. We therefore find ourselves in a position similar to that of someone looking at Van Gogh's famous shoes: there is no object-shoe in the painting, yet the painting unveils the world behind it, the life of the farmer who wore them, the hard dirt where the scuff marks come from. For this to happen, the shoes do not need to exist as objects. In the same vein, painting transitional objects (which are potentially visible, *qua* objects, only to their "owners") can generate an entire web of signifiers that stand for the transitional field within which the relationship with the transitional objects had developed. We neither see nor experience the object (which never exists as object); in place of the absent being, we have—as Heidegger would say[73]—its truth. And the truth of the transitional object (not the object itself, which never appears) that resides in the work of art and is produced for artist and spectator alike—via its *signifiants flottants*—is nothing if not the revelation of the transitional field in which the *couplage* between signifying bodies occurs, between beings that—it bears repeating—are neither objects nor subjects, but σχέσεις, relations.

But let's return for a moment to the exhibition. The transitional objects created for this exhibition are responsible for "photographing" the birth of a transitional object. And even though these products are just representations of the original creation, made in hindsight, they display certain formal and semiotic characteristics that appear to be keeping with the thesis of this article: the transitional object is not merely an imaginative phenomenon, but a semiotic phenomenon (creating and conveying meaning). It is not an object or an epiphenomenon of the subject, but a relationship, σχέσις. As such, the transitional object has a semiotic function similar to a Byzantine icon, as we argue at the end of this article. Let us now say something about these images. In most cases, the transitional object resembles an "emotional trunk," the mythical ancestor of every adolescent diary; an object incapable of telling a story "if it is not allied with other objects." Even when recreated for artistic purposes or for art therapy, the transitional object does not lose its aura of historical authenticity. It can appear as a fragment of past life (a pillow, a blanket, a small album) or as an original construction—as long as there are relationships and atmospheres capable of "physically establishing" that presence which recalls the body image in its historicisation.

Three characteristics seems to be shared by the transitional objects that I have chanced to look at in this collection of works, of which some were by artists and some by ordinary people. These are: miniaturisation, parataxis, and lack of perspective. I do not claim to have exhausted such important issues in a couple of sentences, but I would

---

73 I am referring to the first *Holzwege* by Martin Heidegger, entitled *Der Ursprung des Kunstwerkes* (Frankfurt am Main: Klostermann, 1950), 36–42.

like to offer here a few notes by way of commentary and also propose a brief conclusion to my analysis.

First, the transitional object is often presented as a small world, a miniature version of a larger world. This does not mean that it reproduces the entire external world; it represents a fragment of the world in which the subjects can find some general truth that concerns them, an (un)objective truth. Because it is a reduction, the subjects can appear disproportionately large compared to things. The transitional object is often similar to drawings in which the child depicts himself with a very large head or hands. The miniature is like a synecdoche, only in reverse: it is not the part that stands for the whole, but the whole that seeks to become a part, the subject (always left out of the representation) that lodges itself in the object and makes it concur with itself.

The second characteristic I noted is parataxis. This work on the object becomes work on the subject. What the transitional object constructs is the subject. Like an intransitive verb, it describes an action that takes place within the subject, giving life to the subject itself. In addition, this construction of the subject is unhistorical, set in a time that is always removed from the present experience. Everything happens as if the subject were making his or her transitional object an expressive field devoid of functional parts. There are fragments of things, likenesses, pictures, objects; in other words, a single object, but jagged, partially disjointed, and worn by time. These parts, or this "partial whole," as I call it, is held together without the use of connectives; it is devoid of syntax. The absence of syntax is a characteristic of primitive, oral thought. And in each transitional object it is as if this residual orality is released and takes shape. Partial transitional objects are like Greek epic formulas: they return again and again, and form ritual contexts.

Finally, there is the third feature: perspective. As in folk art, perspective is nowhere to be found. However, whereas in folk art the lack of perspective is a product of improvisation, what we are dealing with in the transitional objects exhibited in Macerata is a poetic choice. Perspective presents us with a centred view of space and time. In these works, in turn, what should be—and is—represented can never be centred, since it coincides with the *space* in which the subject was formed. In short, I am convinced that any artistic transitional object intends speaks to us about the process of world-making; this process is relevant even if the space has not existed forever, but has only begun to exist at a point that remains outside the possibility of representing the subject. The uncentered, unfocused space of the artistic transitional object belies an attempt to give the constitution of the subject the consistency of an object.

If that is true, then transitional objects teach that the subject is made of objects. It is the place and history of the encounter between subjective demands and objective goals—which are different but not irreconcilable. Subject and object, reconsidered in light of the concepts of *objet mamaïsé* and transitional objects, should perhaps be transfigured into a new and perspicuous dimension, where art is no longer an action that produces works, but a model for every construction of the self in the world. In quite similar terms, Yannaras writes that, in studying a painting, it is not the thing that approaches truth, but "the space of personal relation, the immediacy of personal uniqueness and dissimilarity which is experienced vividly in spite of the dimensional non-presence of the person."[74]

Towards a New Paradigm for Human Studies

The time has come to sketch a conclusion. We began with the relationship between subject and object, which we identified as the *locus deperditus* (to borrow an expression from philology) of the present epistemological crisis troubling the human sciences. Because the distinction between subject and object is not evident in nature, nor is it passed down through sensory perception, we focused our attention on semiotic forms. In inevitably summary fashion, we have identified the essential points of the new scientific paradigm, which has replaced traditional humanism and which conditions the understanding of subject and object. The two terms appear stripped of meaning, since the phil-

---

74 Yannaras, *Person and Eros*, 116.

osophical reasoning that heralded their birth and long life has fallen into disuse. Everywhere, in philosophy especially, people speak of the death of the subject and the disappearance of the object.[75] Rather than take a side, I have focused on the intermediary entity, the transitional object, which psychoanalysis introduced in order to clarify ambiguities about the construction of subjectivity and its psychoses. We found the transitional object interesting for several reasons: it is not a real object; it is not a mere extension of subjectivity; it is not a figment of the imagination; it has something of the symbolic and semiotic in it; moreover, it is a web of signifiers (called *objets mamaïsés*) that always permits us to re-create the mother/child relationship (σχέσις) starting with the dyad's sameness of being.

As Simone Weil saw clearly, the new physical and human sciences do not enable us to understand the human as a correlation between the subject and the object of a cognitive representation based on the categories of πρᾶξις or human ποίησις. The expropriation of subjectivity performed by this method affects the metaphysical foundations that until now have upheld or accepted the analytical paradigm. If the familiar subject can no longer be placed at the intersection of the objects onto which it projects its own anthropic image, then there is no point in continuing to call human sciences those protocols of examination based on the subject/object dyad and the cognitive form from which it arises, i.e., Aristotelian logic.

How can the present discussion be of use to the debate about human studies? Two important points have been identified above. First, I have shown that the transitional field upsets the view of subject/object as a dichotomy, on which the common understanding of knowledge is based. By declining to present the humanities as the relationship between an actant subject and an actant object, we can discover relationships previously overlooked—like the adjacency to proximal signifiers

---

75   See the papers collected in the monographic issue *Au-delà du sujet: L'impersonnel?* of *Archives de Philosophie* 76:3 (2013).

that enable us to "recognise ourselves without being ourselves."[76] Second, the transitional object seems to function like an icon: instead of a *Bedeutung*, it requires a web of signifiers. The icon/prototype model of Byzantine iconological semiosis seems to be a valid substitute for the traditional model of conceptual signification, widespread in the West.

The form of signification that we would like to put to the test, in order to build a new model of human and humanities knowledge, is not that of Hjelmslev and his school. It much more closely resembles the generative semiotics of Greimas and Rastier, from which it borrows essential anthropological aspects. But there is one important difference: it explicitly draws on the theory of knowledge developed by Byzantine theologians and on the kind of signification that emerges from the relationship between icon and prototype. The icon, that signifier from the distant past, can help us reimagine our relationship to the human without reducing the human to an object and thereby alienating it and stripping it of historical poignancy. We must go back to the Greek fathers, whose language "functions iconologically."[77] And we must also test the hypothesis of a semiotics that functions iconologically, translating into images those meanings that emerge from anthropic zones. Images are crucial to the new humanism, not only because society at present is ruled by images: that would merely be an ontic fact. Images are *ontologically* crucial, because "the language of images conceals the truth like a dynamic leaven in the mystagogic space of personal relation."[78]

Bidding goodbye to the subject/object dyad need not spell defeat for scholars; on the contrary, it could pave the way toward a different, richer vision of the act of knowledge. The cold, logical space of the oppositional subject/object relationship is replaced with an anthropological space made up of zones of anthropic interaction. The crucial elements are contained in the human ability to establish relationships between the identity and distal zones, to conjure up a transcendent

---

76 I would express my concept by way of the ancient Greek, as follows: ἀλλήλων γνωσθέντων, οὐκ ἀτὰρ ἀλλήλων ὄντων (knowing each other, but not sharing each other's being). That is the very idea of being one another's *alleloi*.
77 Yannaras, *Person and Eros*, 194.
78 Yannaras, *Person and Eros*, 196.

dimension, compared to the basic empiricism of proximal linkages—which is also present in creatures that are not human. The theory of anthropic zones—as modified in what I have proposed here—can offer a useful model for rethinking the general form of knowledge, substituting subject/object categories with categories of proximal and distal linkages that bring the cognitive act closer to a relationship between a proximal signifier and a distal signifier with a hierotopic and liturgical space.

My proposal stands on the shoulders of a few giants; among those, I would like to single out Rastier, who has been a constant source of inspiration. The idea of anthropic zones, which I consider a serious alternative to the traditional concessions of subject/object and subject/predicate logic, is his. Now, if we extend the theory of anthropic zones to the field of the humanities, we can detect a new objective in our investigation. The above analysis has unearthed two new pieces of information. First, we discovered that the dominant epistemology in human sciences aims to represent scientifically a connection between the human subject and the human object. For various reasons, mentioned earlier, the laws governing the new human sciences no longer correspond to the humanistic vision handed down to us from philology; they turn out to be much more similar, in their objectives and methods, to the laws governing the physical sciences. Second, the critical point of this vision was located in the subject/object model, which, applied liberally, produces a vision of the human characterised by the predictability of studied phenomena. The anthropic zones permit us to shed the cumbersome subject/object dyad and its cold logic, and instead adopt an anthropological vision of the space of signification and the signifiers that dwell there. I am proposing a paradigm shift. I am also proposing to rewrite the cognitive model by bidding farewell to the logical subject/object dyad and switching over to an anthropological arrangement of the human zones. In such a design, what is important is understanding how the anthropic zones enable us to conceive of the human not in terms of a contrast between a knowing subject and a known object, but rather as a *couplage* of subjects interacting in a complex space. This means suggesting the shift from a logical vision of the cognitive process to a semiotic-anthropological vision of the human and the social sciences.

To save what is humane in the human sciences, we must abandon the subject/object dichotomy and adopt a semiotic-anthropological view based on the interaction of subjects in constant dialogue with a signifying space that resembles, in its nature, the transitional field in psychoanalysis and—via a specific interpretation of that field—the function of icons in human culture. It also means abandoning the idea that the responsibility of the human sciences is to explain the human being. Instead, their objective should be to clarify humanity's plural and ecological character: not the generic human being, but humans—in all their plurality—inhabit the anthropic zone and, by being in constant dialogue with their *Umwelt*, can render it a shared space capable of evoking distal spaces. Just maybe, it is through merely such a small opening that one day, in the peaceful unconsciousness of time, the Messiah will enter contemporary human and social sciences. Hopefully, the same revolution will also become possible in the hard sciences.

The author reports there are no competing interests to declare.
Received: 24/02/22 Accepted: 12/09/22 Published: 03/10/22

Evolution as History: Phylogenetics of Genomes and Manuscripts

Graeme Finlay[1]

Abstract: The lines of biological evolution are documented in the genomic "texts" of species. Phylogenies of texts, both genetic and literary, can be studied by the same methodologies. In each case, scholars use the presence of variants to elucidate the history of their chosen text—whether it be genetic (the four chemical letters inscribed in DNA) or alphabetic (the letters of biblical languages such as Hebrew and Greek). Several conclusions arise. First, genetic and textual variants constitute the data from which phylogenetic trees of organisms and manuscripts (respectively) may be constructed. Second, such analyses assume the existence of (now extinct) ancestral genomes and ancestral texts, providing evidence that such *urtexts* existed and enable their reconstruction. Third, biological evolution belongs to the category of history, and like all histories, can be understood as development within the created order. Fourth, biological evolution raises questions about divine providence that are similar to questions that arise from any other history. Fifth, theologians need to develop a theology of evolutionary history in the same way as they seek to understand God's

---

Graeme Finlay is retired from teaching scientific pathology at the University of Auckland and is a lay preacher. He has written *Human Evolution* (Cambridge University Press, 2013), *The Gospel According to Dawkins* (Austin-Macauley, 2017), *Evolution and Eschatology* (Wipf and Stock, 2021), and *God's Gift of Science* (Wipf and Stock, 2022). He is married to Jean, a musician, and they have two adult offspring.

1   I thank Dr Liliya Doronina (University of Munster) for permission to use SplitsTree images from her recent publications describing relationships between orders of mammals (footnotes 18 and 19), and Dr Roest Crollius (Institut de Biologie de l'École Normale Supérieure) for karyotype images (note 11). I am grateful to Rev. Dr Jac Perrin (North Central University, Minnesota) and Dr Andrew Edmondson (University of Birmingham) for permission to use phylogenetic images from their studies on biblical textual criticism.

action in biblical history (allowing that only the latter involves personal creatures).

Keywords: comparative genomics; evolution; history; providence; textual criticism

We can often recognise family likenesses. Shared gene variants underlie such resemblances—although shared environments also contribute to similar phenotypes. Genomic data, especially from DNA sequencing, are routinely used to assess the relatedness of individuals and, increasingly, to assess the relatedness of species and the routes by which a single progenitor species can diversify into multiple descendant species.

The phylogenetic development of organisms has been depicted in evolutionary trees for many years. In the genomic era, such trees are constructed using genetic variants. Such powerful genomic markers include chromosome rearrangements and genetic parasites (endogenous retroviruses and transposable elements). These markers outline the transformations of genomes including the route of humanity's evolutionary past.

But some Christians may find phylogenetic trees disconcerting. For the Bible says we are created, and some believers think that a continuous link with monkey forebears contradicts this claim. Phylogenetic trees indicate that diverse living forms are derived from common ancestors by what appear to be mechanistically describable biological processes. Continuous process seems to rule out direct action of a creator God. Miraculous acts of creation become redundant.

This paper will describe how biological evolution (including ours) is demonstrated by comparing variations in genomes of multiple species. The genomic approach will be expounded by comparing it with a topic familiar to Christians—the application of textual criticism (where *criticism* means *analysis*) to describe the history of ancient biblical manuscripts. There is continuity between the genetic texts (genomes) of species as there is between the written texts of manuscript traditions. The evolution of genomes reveals biological history in the same way as the evolution of variants of biblical manuscripts reveals the literary history of those texts. Such progressive transformations are equally *histories*, developments within God's created world. Biological evolution and the biblical concept of creation belong to alternative ontological categories, but they are not opposed to each other. While evolution is a freely operating historical process, it is an ever-dependent one, ordained by God.

Chromosome History: Cutting and Pasting

When cells divide, their DNA and associated proteins are packaged in chromosomes, bodies that can be seen using conventional light microscopes. Using appropriate stains, individual chromosomes, and parts thereof, can be identified. The normal human genome consists of forty-six chromosomes including twenty-two pairs of autosomes and two sex chromosomes (XX in females; XY in males).[2]

Closely related species have similarly structured chromosome sets (or *karyotypes*). By aligning chromosome sets from different species, cytogeneticists can detect differences between them, such as fissions (a chromosome present in one species has split to form two chromosomes in other species), fusions (two chromosomes have joined end-to-end) and reciprocal translocations (chromosomes have exchanged lengths of material). More subtle, but more frequent, rearrangements include inversions (a block of chromosomal material flips 180° with respect to its surrounds), duplications, insertions, and deletions. Stepwise chromosomal rearrangements are familiar natural phenomena, and frequent in cancers.

The alignment of karyotypes from related species demonstrates how one karyotype can be transformed into another by cutting-and-pasting chromosomal material. In addition, such comparative studies allow reconstruction of the karyotypes of common ancestors (now extinct). We could say that we may infer the genomic

---

2   There is also a tiny chromosome in mitochondria, but this is not relevant to the current discussion.

urtext, the ancestral version from which all derivative extant genome arrangements are derived. In biblical textual criticism, the *urtext* is the original form of a composition. We could call the ancestral karyotype of a group of species its *urkaryotype*.

The lines of primate chromosome evolution have been delineated by this microscopic level comparative cytogenetics approach.[3] More remotely, the ancestral eutherian[4] karyotype has been reconstructed, and can be transformed into the ancestral primate karyotype by three fissions and two fusions. The latter can be transformed into the ancestral anthropoid (simian; monkey-ape) karyotype by a fission, a fusion, and a translocation, and from thence into the ancestral hominoid (ape, including human) karyotype by another fission. Morphological comparisons indicate that the hominoid common ancestor had 48 chromosomes. Humans have 46 chromosomes because two chromosomes retained in the chimpanzee genome fused to form human chromosome 2 (via an inversion in each of the two precursor chromosomes that occurred in a human-chimpanzee ancestor).

These insights, obtained by observing chromosomes microscopically, are of relatively low resolution. The era of high-throughput genome sequencing (with computational analysis of sequences) now allows chromosome rearrangements to be identified and mapped at high resolution. Genome evolution can be studied at the level of the genetic text.

The karyotype of the eutherian ancestor, that lived at least 80 million years ago,[5] can be rearranged into the human karyotype by 162 DNA breakage events (of which the most common generated in-

---

[3] Roscoe R. Stanyon, Mariano Rocchi, Oronzo Capozzi et al., "Primate Chromosome Evolution: Ancestral Karyotypes, Marker Order and Neocentromeres," *Chromosome Research* 16 (2008): 17–39, DOI: 10.1007/s10577-007-1209-z; Steffan Muller, "Primate Chromosome Evolution," in *Genomic Disorders: The Genomic Basis of Disease*, ed. James R. Lupski and Pawel Stankiewicz (Totowa: Humana, 2006), 133–152, DOI: 10.1007/978-1-59745-039-3_9.

[4] Eutherian mammals have a well-developed placenta and include all extant mammals except for monotremes (such as platypus) and marsupials.

[5] Sandra Alvarez-Carretero, Asif U. Tamuri, Matteo Battini et al., "A species-level timeline of mammal evolution integrating phylogenomic data," *Nature* 602 (2022): 263–267, DOI: 10.1038/s41586-021-04341-1.

versions).[6] The karyotypes of long-extinct ancestors intermediate between the eutherian and human have also been deduced from the chromosome complements of extant species. Figure 1 depicts the number of inferred DNA breakages that converted the karyotype of a eutherian ancestor into the human one.

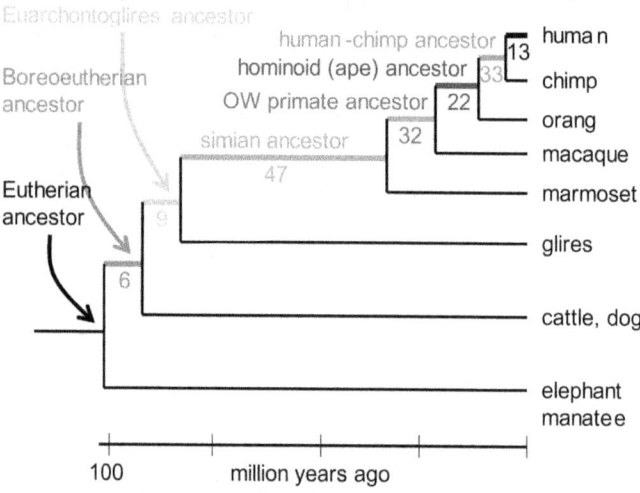

Figure 1. DNA breakpoints reconstructed from comparing karyotypes of extant species

The resolution of analysis was DNA segments ≥300,000 bases long. Glires include rodents and rabbits. Boreoeutherians are a major category of mammals, that include primates, rodents, hoofed mammals and whales, carnivores, and bats. Adapted from Kim et al. (2017), note 6.

More recently, genomes of species yet more distant from ours have been sequenced. Alignments including the platypus and echidna genomes (along with karyotypes of marsupials, chicken, and a lizard) have enabled geneticists to reconstruct the karyotype of the *mamma-*

---

[6] Kim Jaebum, Marta Farre, Loretta Auvil et al., "Reconstruction and evolutionary history of eutherian chromosomes," *Proceedings of the National Academy of Sciences of the USA* 114 (2017): e5379–88, DOI: 10.1073/pnas.1702012114.

lian ancestor. Some 165 rearrangements were needed to convert the ancestral mammalian genome, the mammalian genetic *urtext*, into the human one.[7]

Similar reconstructions based on the karyotypes of extant birds, have indicated the likely karyotypes of a bird ancestor, of a bird-turtle (archelosaur) ancestor, and of a bird-reptile (diapsid) ancestor. As with the mammal karyotype, the chromosome sets are interconvertable by the chromosome rearrangements familiar to geneticists. Seven fissions are needed to convert the turtle karyotype into the avian one.[8]

Karyotypes of birds, of turtles, and of squamates (snakes, lizards) include tiny *microchromosomes* as well as conventional macrochromosomes. Certain microchromosomes in different species contain the same genes and must have been inherited intact from the same ancestor. They are related also to the microchromosomes which comprise the genome of amphioxus, a little fish-like creature that is classified near the base of the chordate family tree.[9] In birds and reptiles, the number of microchromosomes tends to decrease due to fusions that generate macrochromosomes.[10] Platypuses have several small chromosomes that are products of micro-micro fusions. All other mammals lack microchromosomes, and segments derived from microchromosomes have been fragmented beyond recognition by genome rearrangements.

---

7   Yang Zhou, Linda Shearwin-Whyatt, Jing Li et al., "Platypus and Echidna Genomes Reveal Mammalian Biology and Evolution," *Nature* 592 (2021): 756–762, DOI: 10.1038/s41586-020-03039-0.
8   Darren K. Griffin, Denis M. Larkin, and Rebecca E. O'Conner, "Time Lapse: A Glimpse into Prehistoric Genomics," *European Journal of Medical Genetics* 63 (2020): 103640, DOI: 10.1016/j.ejmg.2019.03.004.
9   More correctly, amphioxus is a sister group to vertebrates. The vertebrate-amphioxus divide has been dated to 684 million years ago.
10 Paul D. Waters, Hardip R. Patel, Aurora Ruiz-Herrera et al., "Microchromosomes are Building Blocks of Bird, Reptile, and Mammal Chromosomes," *Proceedings of the National Academy of Sciences of the USA* 118 (2021): e2112494118, DOI: 10.1073/pnas.2112494118.

An ever-increasing database of sequenced genomes, coupled with sophisticated algorithms, is facilitating the reconstruction (at higher resolution) of even more distant ancestral karyotypes. The number of reconstructed ancestral genomes approximates the number of sequenced extant genomes.[11] An example, shown in Figure 2, depicts how a reconstructed genomic arrangement belonging to an ape ancestor can be rearranged to form the human and (more fragmented) gibbon karyotypes.[12]

An urkaryotype of an ancestor of all multicellular animals may be derived by comparative analysis of more simple animals. Organisms involved in this work included sponges, cnidarians (sea anemones, corals, jellyfish) and bilaterians (worms, molluscs, and chordates).[13]

DNA is chemically simple but informationally rich—a text of extraordinarily dense content. It contains a detailed record of a species' history. Alignments of the genetic texts belonging to multiple related species or groups of species reveal when novelties appeared. When increasingly remotely related species are used in the intertextual comparisons, a series of ancestral karyotypes can be inferred.

---

[11] Nga Thi Thuy Nguyen, Pierre Vincens, Jean Francois Dufayard et al., "Genomicus in 2022: Comparative Tools for Thousands of Genomes and Reconstructed Ancestors," *Nucleic Acids Research* 50 (2022): D1025-31, DOI: 10.1093/nar/gkab1091; Matthieu Muffato, Alexandra Louis, Nga Thi Thuy Nguyen et al., "Reconstruction of Hundreds of Reference Ancestral Genomes across the Eukaryotic Kingdom," *Nature Ecology and Evolution* 7 (2023): 355–366, DOI: 10.1038/s41559-022-01956-z.

[12] A detailed presentation of karyotypic evolution in higher primates is provided at the wonderfully interactive and illuminating website: https://www.genomicus.bio.ens.psl.eu/genomicus-102.01/cgi-bin/karyotype_handle.pl? numChrom_ef=30&numChrom=30&minGene=50&display=137%3A409%3A413%3A4 14%3A149%3A150%3A151%3A152%3A&reverse=149&species_id=137

[13] Oleg Simakov, Jessen Bredesen, Kodiac Berkoff et al., "Deeply Conserved Synteny and the Evolution of Metazoan Chromosomes," *Science Advances* 8 (2022): eabi5884, DOI: 10.1126/sciadv.abi5884.

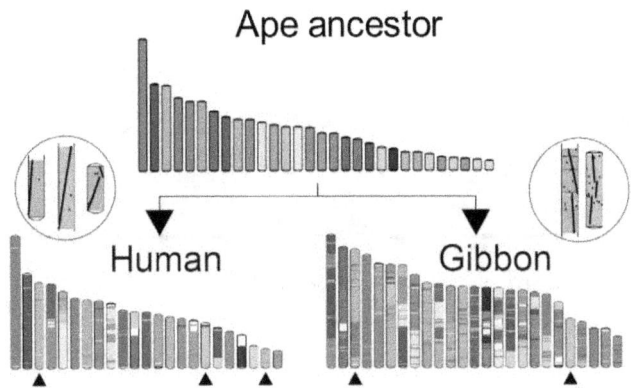

Figure 2. Cutting-and-pasting chromosomes: converting the ancestral ape karyotype into the human and gibbon karyotypes

The inferred ancestral ape (hominoid) karyotypic arrangement is shown. Each coloured bar represents a block of chromosome material (a *contiguous ancestral region*). Chromosomal blocks are colour-coded to indicate relation to the ancestral regions. The human karyotype is less rearranged than that of gibbon. From Muffato et al. (2022), note 11. Used with kind permission of Dr Roest Crollius.

Parasites in Our Genome

Genomes are not static assemblages of genes. New genetic material is constantly added. One source of novel DNA is a category of infectious agents called *retroviruses*. When a cell is infected, the retroviral genome is copied from RNA into DNA by a retrovirus-encoded enzyme called a *reverse transcriptase*. Another viral enzyme, an *endonuclease*, selects (at random) a point in the host DNA (the *target site*), and splices the retroviral DNA into the host DNA at this site. During this process, the target site is duplicated so as to bracket the new segment of retroviral DNA. If the infected cell is a *reproductive* cell that can transmit its DNA to future generations, the inserted retroviral DNA may become a feature of the genome of the species. It is said to be *endogenous*. This process is depicted in Figure 3 (left scheme).

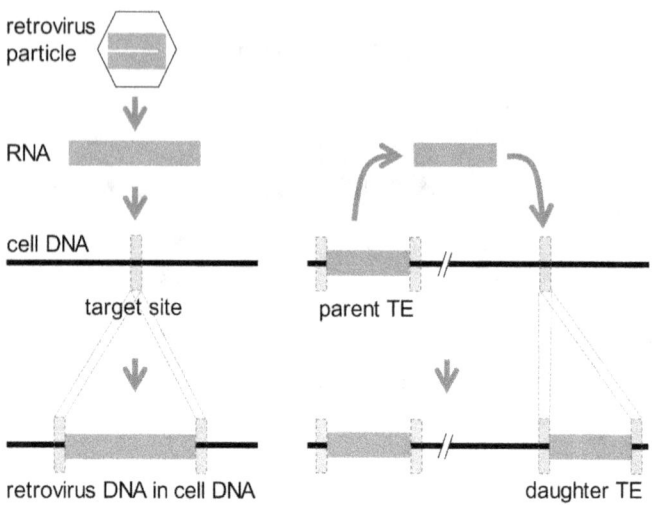

Figure 3. Agents that modify genomes

Most of our genome is comprised of parasitic units of DNA. These are of two main types.

Left: Retroviruses (hexagon) colonize DNA by introducing their RNA (green bar) into chromosomal DNA of infected cells. Viral enzymes select a target site (red dotted box) in the cell DNA at which the retroviral genome is copied into DNA and inserted into the cell's DNA. During this process, the target site is duplicated.

Right: Transposable elements (TEs) are of many types, but typically reproduce in genomes when a parent TE is copied into RNA. Enzymes produced by TEs select a target site elsewhere in the genome (red dotted box) at which the TE RNA is copied into DNA and inserted into the cell's DNA. Again, the target site is duplicated.

Another broad category of parasitic DNA resides in and colonises genomes. There are hundreds of different subtypes of such parasitic DNAs in genomes such as ours, and collectively they are known as *transposable elements* (TEs). Most of these multiply by a copy-and-paste strategy using TE-encoded reverse transcriptase/endonuclease enzymes. They also generate target site duplications as they replicate (Figure 3, right scheme). As with retroviruses, if a new TE is generated in a reproductive cell, it may be passed on to future generations and become a characteristic feature of a species' genome.

Endogenous retroviruses and TEs together comprise more than fifty percent of our entire genome. The vast majority of these are shared by all human beings. The question arises as to when they entered our genomes. As with studies on chromosome number and structure (karyotype), we can best address this question by aligning the genome sequences of human and other species and ascertaining whether a particular element is shared by multiple species. Such a genomic comparison is exemplified for a TE called an SVA element that is located near the *SHPK* gene (Figure 4).[14] The sequence of the four letters (A, C, G, T) that comprise the genetic "text" around the insertion site is presented.

This alignment shows that the SVA element entered primate DNA in an ancestor of the African great apes (human, chimp, bonobo, gorilla). The undisturbed target site is present in Asian apes (orangutan, gibbon), Old World monkeys (four species shown), and New World monkeys (four species).

SVA elements appeared only in great apes. They were cobbled together from bits of genetic flotsam. They are mutagenic (alter DNA sequences) and certain insertions (depending on their genomic location) cause genetic diseases. Thousands of SVA elements are present in great ape genomes, and the host species of each element is known.[15] Five hundred of these have been selected to create a phylogenetic tree of the great apes (Figure 5). Most SVA elements have arisen relatively recently and are found only in one species. For example, 98 are found only in the human genome. Forty-four inserts are shared by human and chimp genomes. They were added to the genomes of human-chimp ancestors.

14 Emma Price, Olympia Gianfrancesco, Patrick T. Harrison et al., "CRISPR Deletion of a SVA Retrotransposon Demonstrates Function as a cis-Regulatory Element at the TRPV1/TRPV3 Intergenic Region," *International Journal of Molecular Sciences* 22 (2021): 1911, DOI: 10.3390/ijms22041911.
15 Orr Levy, Binyamin A. Knisbacher, Erez Y. Levanon, and Shlomo Havlin, "Integrating Networks and Comparative Genomics Reveals Retroelement Proliferation Dynamics in Hominid Genomes," *Science Advances* 3 (2017): e1701256, DOI: 10.1126/sciadv.1701256.

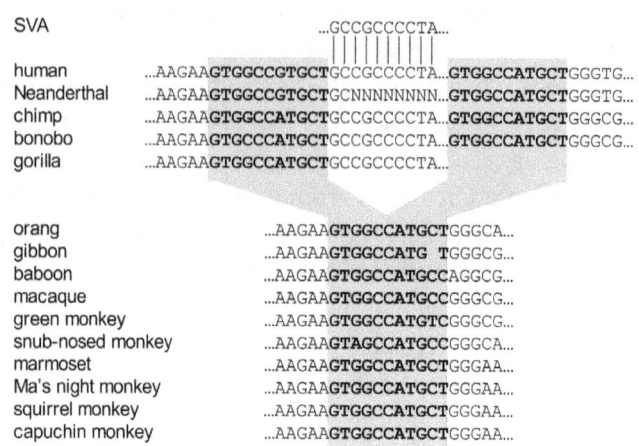

Figure 4. The insertion site of an SVA element in humans and other primate species.

The DNA sequence alignment represents a comparison of genetic text that indicates when in primate history the insertion mutation occurred. The SVA element (starting GCCGCCCTA...) is present between duplicates of the target site (in bold text and shaded) in humans, chimps, bonobos, and gorillas. (In the case of gorillas, right-hand sequences have been deleted). Other species retain the original undisturbed target site. In the human and Neanderthal left-hand target site duplicate, the seventh based has mutated to a G, whereas the same position in all other cases is A. A sequence gap is indicated by "N." This insert is near the *SHPK* (sedoheptulokinase) gene. From Price et al. (2021), note 14.

Ninety are shared by human, chimp, and gorilla genomes. Each one of these is a powerful demonstration of African great ape monophylicity (descent from the same ancestral linage). And a handful are common to all great ape genomes (but were not present in the selection of inserts depicted in Figure 5).

Somewhat mysteriously, five SVA inserts are present in human and gorilla genomes, but not that of chimpanzees (Figure 5, "HG, 5"). At first glance, this is not compatible with the phylogenetic tree. We can account for anomalous insertions by a phenomenon called *incomplete lineage sorting*. Some anomalous inserts are inevitable, as illustrated in Figure 5.

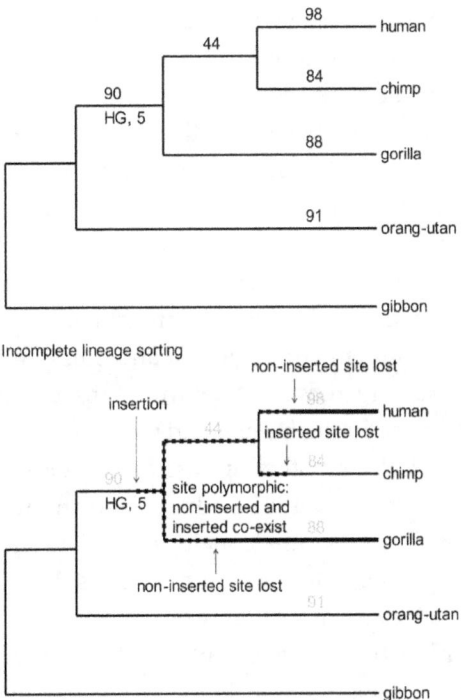

Figure 5. A phylogenetic tree of great apes based on SVA element insertions

Above: In the random selection of SVA elements used above, none was present in humans, chimps, gorillas, and orangutans. However, several have been documented, establishing great ape monophylicity. From Levy et al. (2017), note 15, Data fileS1, SupData3 (orthologues table), with 7488 individual SVA insertions.

Below: *Incomplete lineage sorting* occurs when a retrovirus or TE is inserted into germline DNA near the time when species diverge. The original chromosome without the insert, and the derived chromosome with it, will coexist in the population. The site is *polymorphic* with respect to the inserted element. As the nascent species develop, the chromosome with the insert may randomly drift in frequency so that it may be either lost (only the pre-insertion chromosome is retained) or *fixed* (the pre-insertion chromosome is lost).

Every event involving the insertion of a transposable element is unique. The new element may be transmitted to members of the family, then to members of the tribe, and eventually to individuals in the wider population. For a considerable period, both the original, pre-inserted sequence and the derived SVA element-containing sequence will be present in the genetic pool of the population. The site is said to be *polymorphic*—it has two alternative sequence features. If two or more species diverge at this point, the element will be inherited in a polymorphic state in all nascent species. As a result of random drift, it will eventually either be lost, or it will displace the original pre-inserted sequence.[16] In the latter case the new element becomes *fixed* as a property of the genome. The five SVA inserts noted in Figure 5 arose in the genomes of human-chimp-gorilla ancestors, were polymorphic when the species diverged, were lost during subsequent chimp history, but retained and fixed in the human and gorilla lineages.

TEs have featured in this discussion because they have had a huge effect on genome evolution and are powerful markers of phylogenetic relationships. The same conclusions are reached by comparative studies using other categories of mutations. For example, simple insertions and deletions in genomes of multiple species corroborate the accepted outline of primate evolution. These uniquely arising mutations demonstrate the monophylicity of the African great apes, and the fact that humans are more closely related to chimps than they are to gorillas.[17]

Incomplete lineage sorting is rampant during bursts of speciation. A mammalian superorder called Euarchontoglires includes five orders: primates, flying lemurs, tree shrews, rodents, and rabbits. Selection of a subset of transposable elements that proliferated

---

16   In general, over evolutionary timescales, the only stable frequencies of a genetic variant are 0% (the element is lost) or 100% (the element is fixed). This assumes that the element drifts in frequency in a random way.

17   James K. Schull, Yatish Turakhia, James A. Hemker et al., "Champagne: Automated Whole-Genome Phylogenomic Character Matrix Method Using Large Genomic Indels for Homoplasy-Free Inference," *Genome Biology and Evolution* 14 (2022): evac013, DOI: 10.1093/gbe/evac013.

during early Euarchontoglires history shows complex patterns of element presence and absence, and of phylogenetic relationships.[18]

Similarly, Laurasiatherian mammals include moles, cattle and whales, carnivores, pangolins, horses and bats. Incomplete lineage sorting was rife in the early days of their evolution. A group of informative transposable elements indicates that these orders diverged from the Laurasiatherian ancestral species in complex ways.[19]

A standard phylogenetic tree (such as those of Figures 1 and 5) cannot depict the network of relationships entailed in widespread incomplete lineage sorting. Mathematicians have devised computational methods that can express the complexities of anomalous trees. An informative analysis is that of the SplitsTrees algorithm, which presents the early phylogenetic relationships of diverging species as networks (Figure 6). These show (for example) that primates are most closely related to flying lemurs, followed by tree shrews; and that carnivores are most closely related to pangolins and then to cattle and whales.

Phylogenetic Trees Reveal the Shape of History: Comparison with Ancient Manuscripts

The era of comparative genomics has facilitated the application of genetic "textual criticism"—the analysis of changes in the DNA "text" that have occurred during evolution. Biblical textual critics have been doing the same thing with ancient manuscripts for many years. In 1832, Lachmann proposed that manuscripts that share common errors (especially highly distinctive ones) have a common ancestry. This is because the presence of an "indicative error" in two or more manuscripts

---

18   Liliya Doronina, Olga Reising, Hiram Clawson et al., "Euarchontoglires Challenged by Incomplete Lineage Sorting," *Genes* 13 (2022): 774, DOI: 10.3390/genes13050774.

19   Liliya Doronina, Graham M. Hughes, Diana Moreno-Santillan et al., "Contradictory Phylogenetic Signals in the Laurasiatheria Anomaly Zone," *Genes* 13 (2022): 766, DOI: 10.3390/genes13050766.

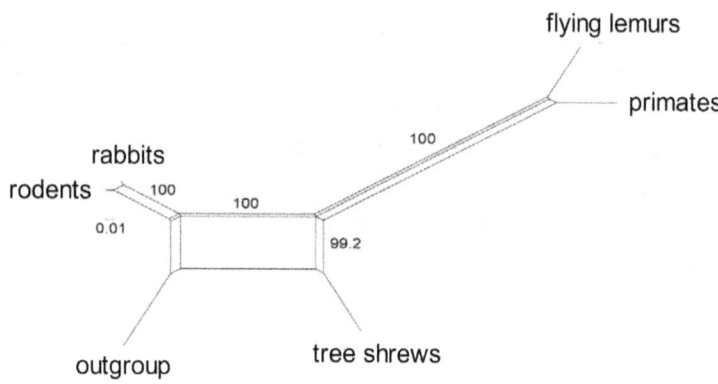

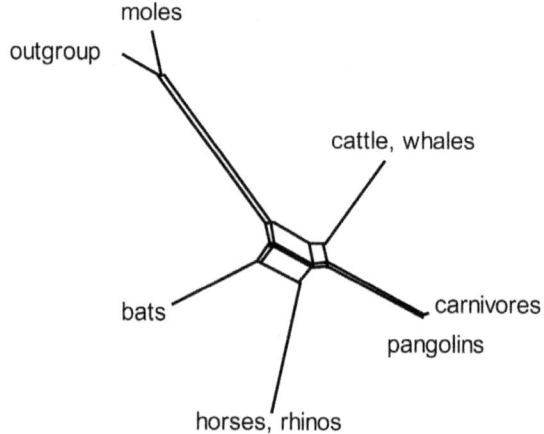

Figure 6. Incomplete lineage sorting during mammalian evolution: SplitsTrees representation

The SplitsTree diagram shows the relationships of the orders included in Euarchontoglires (above) and six orders in Laurasiatheria. It is based on the presence or absence of a set of transposable elements. Both groups show a network of connections during the early burst of speciation. From Doronina, Reising et al. (2022), note 18, and Doronina, Hughes et al. (2022), note 19.

cannot have been made on two separate occasions.[20] This information may be used to construct family trees (*stemmata*) of manuscripts.[21] Variants in extant manuscripts can reveal when those novelties arose in (now lost) ancestral manuscripts. These concepts are expanded below.

The genome undergoes several mutations every time a cell replicates its DNA in preparation for cell division. Ancient manuscripts generated by manual copyists undergo textual changes every time a manuscript is copied. In each case, new variants may be preserved in succeeding iterations of the copying process. Genomic texts and ancient written biblical texts thus accumulate mutations or variants with successive copying. Such variants act as markers that can be used to trace the history of the text. The phylogenetics of species and of ancient texts are closely analogous processes.[22] Indeed, the New Testament scholars Wright and Bird speak of "the living process of textual transmission."[23] Textual difference could be understood "as a stage in a living text's adaptation to its environment."[24]

The approach of constructing phylogenetic trees or stemmata using genomes or literary texts requires that the researcher *aligns texts* from various sources, and *notes variants* present in them. Some variants will be peculiar to a single text, and others may be represented in multiple different texts, suggesting that those many texts are related.

Genetic processes that have analogies with copying errors during textual transmission include recombination (a scribe may switch copying from one manuscript to another), convergence (scribes may independently introduce the same changes that reflect, for example, local dialect) and transposition (!) (when text from one passage is inserted into another). Phylogenetic analyses can be confounded by reversion

---

20 Any such error occurred *once*, and its presence in more than one manuscript demonstrates that it was propagated by *copying*. This is the same logic whereby the presence of "homoplasy-free" mutations (such as ERV and TE inserts) is taken to establish that multiple species have a common ancestor.
21 Andrew C. Edmondson, "An Analysis of the Coherence-Based Genealogical Method Using Phylogenetics," PhD Diss. (University of Birmingham, 2018), 166.
22 Edmondson, "Analysis," 171–173.
23 N. T. Wright and Michael Bird, *The New Testament in Its World* (London: SPCK, 2019), 853.
24 Yii-Jan Lin, in Edmondson, "Analysis," 171.

of a mutation, which is an analogous process to the correction of an error by a copyist.[25]

Based on the occurrence of variants, the next step is to *group texts* (genomic and literary) with the same variants into families. In both cases, deviation from an ancestral text becomes accentuated with the number of times a text is copied. Biological organisms show increasingly marked genetic divergence as family connectedness becomes more remote: from populations within a species to closely related species (genera), families, orders, and classes of species. In the same way, manuscripts differ from each other and textual critics use metaphors such as family, clan, and tribe to categorise them into related groups.[26]

DNA and textual variants may be *classified* as progenitor or derived. For example, if a variant X in one or more examples is present only when variant Y is also present, but Y can be present without X, then it appears that Y existed before X, which arose in a text already possessing Y.

A-B-C-D-E-F-G the standard reading in a set of texts
A-B-C-Y-E-F-G population of texts with first mutation Y
A-B-C-Y-E-X-G texts with a second mutation X appearing as a subset in Y

A long-term purpose is the *reconstruction* of an ancestral text, the progenitor of all the texts which share a set of variants, whether genetic (Figures 2 and 4) or literary. The original text giving rise to a family of texts almost certainly no longer exists. In the language of New Testament scholars, the ancestral text might be the "earliest attainable version,"[27] or the *vorlage* (prototype or template)[28] of a group of texts. Ultimately, it may be possible to reconstruct the *ausgangstext* (the most recent common ancestor) of them all.[29] The original from the writer's

---

25   Howe et al., in Edmondson, "Analysis," 171–172.
26   Jac D. Perrin Jr, "Family 13 in Saint John's Gospel," PhD Diss. (University of Birmingham, 2012), 11.
27   Perrin, "Family 13," 10.
28   Perrin, "Family 13," 17.
29   Edmondson, "Analysis," 13; Wright and Bird, *The New Testament*, 854.

hand (autograph) may not be reconstructible, but the ultimate aim is to approach it as closely as possible. In the case of biblical textual criticism, "the evidence for the New Testament as a whole is massively strong, and we can be quite sure that, despite lots of small-scale variations here and there, we are reading substantially what the writers intended us to read."[30]

The ordering of genetic variants enables derivation of a family (phylogenetic) tree (or *stemma* as a textual critic would say) as exemplified in Figures 1 and 5. An early application of phylogenetic algorithms to literary texts illuminated the history of manuscripts of Chaucer's *Canterbury Tales*.[31] The same algorithms have been applied to biblical manuscripts by Jac Perrin and Andrew Edmondson (references cited). An outcome of phylogenetic analysis of old manuscripts of John's Gospel ("Family 13") is depicted in Figure 7. This analysis identifies ten ancient manuscripts in Family 13, which fall into three colour-coded subgroups. The distinctives of Family 13 texts are indicated by comparison with an outgroup (Erasmus' *Textus Receptus*, TR, red).

There are obvious parallels with family trees constructed from karyotype changes or SVA insertions in hominoids. But a simple tree may not provide all the information needed to relate a group of texts. A hybrid manuscript may be produced from several progenitor manuscript traditions. In this case, the SplitsTrees representation is able to depict the mixing of textual material, as in the analysis of Family 13 (Figure 8). The same subgroups are present, but the network of relationships indicates that the texts were not transmitted in a strictly linear way. Copyists could introduce a variant based on the memory of a manuscript that is different from the one before them. Or they could switch template manuscripts, so that their new manuscript is a hybrid of two or more precursors. We could call this "textual incomplete lineage sorting." A closer biological analogy might be *introgression*, the phenomenon by which two emerging species exchange genetic in-

---

30   Wright and Bird, *The New Testament in its World*, 851.
31   Adrian C. Barbrook, Christopher J. Howe, Norman Blake, and Peter Robinson, "The phylogeny of The Canterbury Tales," *Nature* 394 (1998): 839, https://doi.org/10.1038/29667.

formation through intermittent mating.³² Evolution of species and the development of textual traditions are not strictly linear.

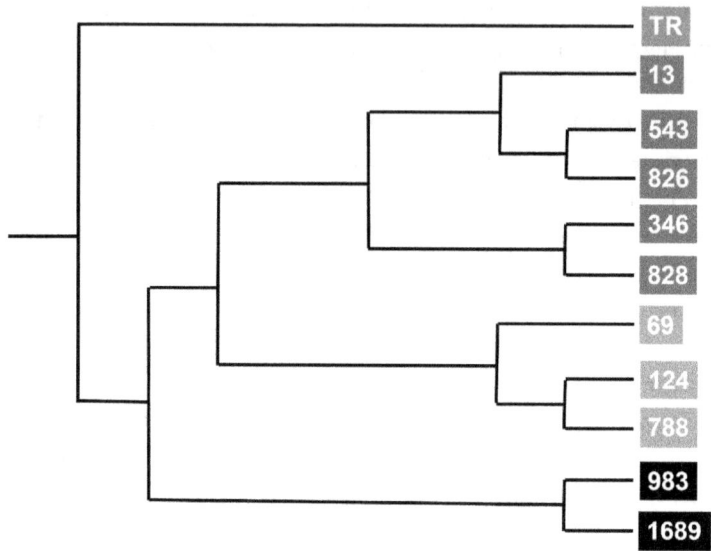

Figure 7. Relationships of Family 13 manuscripts of John's Gospel (PAUP* programme)

Phylogeny of ten manuscripts of John's Gospel that belong to Family 13. A standard text (the *Textus Receptus*, TR; red box) provides a point of comparison, so that characteristics peculiar to Family 13 can be identified. TR provides an *outgroup* that allows the tree to be *rooted* at a particular point. From Perrin, "Family 13," Figure 62 (https://etheses.bham.ac.uk/id/eprint/4482/), and used with permission of the author.

The study of genetic texts is analogous to that of handwritten texts because each presupposes a history. As Perrin has stated, "Ancient manuscripts do not appear ex nihilo"³³—that is, as if by miracle, or instantaneously by divine fiat. Every artefact is part of a continuum of

---

32   Two populations which are in the process of diverging as new species may undergo backcrossing to form hybrids, with genetic admixture.
33   Perrin, "Family 13," 11.

transmission and can be defined by its evolution and context.[34] In the same way, the myriad genomes (genetic texts) that can be aligned with ours show that they have not appeared *ex nihilo*. Genomes including ours have developed, by familiar mechanisms, from those of myriad generations of precedents. Genetic analysis is by its nature the deciphering of history.

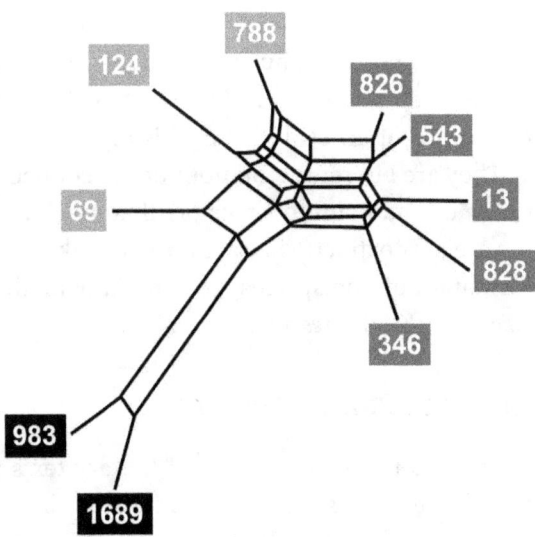

Figure 8. Relationships of Family 13 manuscripts of John's Gospel (SplitsTrees programme)

Phylogeny of ten manuscripts of John's Gospel that belong to Family 13. The *Textus Receptus* provides a point of comparison but is not shown. Modified from Perrin, "Family 13 in Saint John's Gospel," Figure 63 (https://etheses.bham.ac.uk/id/eprint/4482/), and used with permission of the author.

---

34   Perrin, "Family 13," 10.

Genomics reflects textual criticism in another way. Texts can be aligned on the presumption that there is a progenitor *urtext* from which a family of texts was descended. The fact that we can align genomes thus invites the expectation that there is a genomic *urtext* from which a group of extant genomes is descended. We could in principle reconstruct a hominoid (ape) *urtext* belonging to an extinct ape ancestor; or a simian (monkey-ape) *urtext* or, given a sufficient number of genomes, a primate or mammal or amniote (reptile-bird-mammal) *urtext*. As described, this reconstruction is well underway with the analysis of karyotypes and of genomes colonised by transposable elements.

Christians need not fear phylogenetic trees. Those constructed from genomes, including our own, are cogent evidence that our genomes are the record of an evolutionary history that is shared with other species. They are but representations of our connectedness with other extant species and point to ancestors that we share with them. Such ancestors are now extinct. They are missing links but are integral to the transformation of one species' genome into another. But such *histories* are aspects of God's *creation*.

Comparisons with Biblical History

The phylogenetic histories of genomic and literary texts do however differ in one fundamental respect. The former story includes generative trajectories of increasing novelty, complexity, and cognitive capacity. The latter is a degenerative story of progressive alteration of ancestral texts. We must look elsewhere to understand the special nature of genomic history.

Our biological history has analogies with human histories, including those of Israel in the Hebrew Scriptures (or the Christian Old Testament) and of Jesus and his church, as described in the New Testament. There are also differences in these histories. Phylogenetic history has no *personal* content. There is nothing normative, no moral vision, no intimation of the love or goodness of God, no claim upon our loyalties. And, as read through Christian eyes, Old Testament history points to the history of Jesus—incarnation, mission, resurrection—in which the new creation is inaugurated, and the constitution of reality transformed. "Christ is both the Lord of the whole of the history of created reality and the destiny to which all creation is moving."[35] In phylogenetics, as in the biblical portrayal of Israel and the church, we are dealing with phases of history.

Biblical faith is irreducibly historical. This character should have given us the *a priori* expectation that the cosmos and, in our context, biology should also be historical. John Polkinghorne said that it was one of the great discoveries of the twentieth century "that the universe itself has a history and partakes of becoming."[36] Evolutionary history then presents no challenge to faith in the God who works in history.

All histories are *interpreted*.[37] The history of Jesus provides the hermeneutic key by which Christians interpret Israel's history—and the key that enables us to interpret the Primal Testament, evolutionary history as recorded in our DNA, as a part of God's overarching plan for the world. To St Paul, Christ "is the key that opens all the hidden treasures of wisdom and knowledge."[38] The history of Jesus brings intelligibility to the histories of the cosmos, of biology, and of Israel. We may see significance, purpose, and hope in the development of our genome that could never be read from the four chemical bases alone.

History is *continuous*. When the Hebrew Scriptures speak of God's creation of the cosmos, they often use the participial form, indicating God's continuing action.[39] This has been called *creatio continua*.[40] In our spacetime cosmos, time is itself created. From our perspective as players in that cosmos, the continual flow of time—that is, history at any scale or of any entity—is sustained by God. Whether we are looking

35   Adrio Konig, *The Eclipse of Christ in Eschatology* (Blackwood, South Australia: New Creation, 2007), 31.
36   John Polkinghorne, *Science and Creation* (London: SPCK,1988), 39.
37   N. T. Wright, *The New Testament and the People of God* (Minneapolis: Fortress, 1992), 88.
38   Col 2:3.
39   Walter Brueggemann, *Theology of the Old Testament* (Minneapolis: Fortress, 1997), 146, 152.
40   For example, John Polkinghorne, *Science and Christian Belief* (London: SPCK, 1994), 75-76.

at the apparently unchanging stars, the development of vertebrate genomes, or a new baby, all are continually given existence by God.

The physical structure of the cosmos is so constituted as to sustain a history (in all its ambiguities) that will lead to the ultimate purposes of God for a redeemed and transformed humanity.

> God did what he had purposed and made known to us the secret plan he had already decided to complete by means of Christ. This plan, which God will complete when the time is right, is to bring all creation together, everything in heaven and on earth, with Christ as head.[41]

History is *lawful*. Both physical and moral cause-and-effect patterns of order are embedded in the universe. To Douglas Spanner, the cosmos is so constituted that its lawful physical processes, as upheld by God, give to nature "a certain built-in autonomy."[42] Christopher Kaiser has described how the biblical concept of nature's *relative autonomy* facilitated the development of science. Nature is self-sufficient because God has granted it laws of operation.[43] God sustains nature in total faithfulness but grants freedom to the creatures (whether atoms, transposable elements, or people) to behave in the ways consistent with their nature.

The relative autonomy of nature reflects the giving of its own order and laws, and the freedom of the cosmos to evolve in accordance with those constraints.[44] The creator sets the parameters, which describe a fruitful universe with the potential to fulfil his purposes of love. At the same time, the creatures are given freedom of action within those limits.

Brueggemann argues that moral law is built into creation. God's purposes, as given in the Law at Sinai, "are assured in the very fabric

---

41 Eph 1:10–11, GNT.
42 Douglas Spanner, *Biblical Creation and the Theory of Evolution* (Exeter: Paternoster, 1987), 40.
43 Christopher Kaiser, *Creation and the History of Science* (London: Marshal Pickering, 1991), 15–34.
44 As propounded no later than John Philoponus in the sixth century. See Harold Turner, *The Roots of Science* (Auckland: DeepSight Trust, 1998), 101.

of creation."[45] Keeping the commandments is needed for the viability of creation. God's commandments "are not social conveniences or conventional rules." They are "the insistences whereby life in the world is made possible."[46]

History is *contingent*. Peter Harrison has stressed the importance of recognising biological evolution as history.[47] Many nineteenth century Christians were perturbed by Darwin's theory because they thought that the adaptations of organisms demonstrated the elegance of one-off design events. However, if they had only applied God's action in Israel's history to God's action in biological history, they would have seen the same meandering patterns and ambiguous outcomes. Israel's history was contingent, replete with failure, suffering, and calamity, as well as possessing climaxes of beauty. Our biological history also is contingent, with extinction, disease, and predation as well as much to inspire wonder and praise. The Christian belief that natural history and human history are deeply purposive processes is based on revelation: "what made the case for purpose in history was not a case of logical inferences from available facts, but the revealed tradition contained in the Hebrew Bible and the New Testament."[48]

Histories show repeating patterns, *convergence*. In biology, the question has been asked as to what creatures might be generated if the tape of evolution was rerun. Would they be totally different from those that now populate our planet, or would they be similar? Simon Conway Morris has argued that "the number of evolutionary endpoints is limited: by no means everything is possible"; and "what is possible has usually been arrived at multiple times, meaning that the emergence of the various biological properties is effectively inevitable."[49] As three senior physicists state, "Although individual steps in

---

45 Brueggemann, *Theology*, 303.
46 Brueggemann, *Theology*, 201.
47 Peter Harrison, "Evolution, Providence and the Problem of Chance," in *Abraham's Dice*, ed. Karl W. Giberson (Oxford University Press, 2016), 260–290.
48 Harrison, "Evolution," 279.
49 Simon Conway Morris, *Life's Solution: Inevitable Humans in a Lonely Universe* (Cambridge University Press, 2003), xii–xiii.

evolution may be random, the overall direction is constrained by the way the world is."[50]

The same principle seems to hold within the history of human affairs. Free moral choices and visions tend towards certain outcomes. Nick Spencer has suggested that "were we to re-run the tape of Western history, erasing what actually happened and letting it run again, we might, assuming the same deep Christian conditions and commitments, end up with a set of values that, while superficially different, bore a striking resemblance to those we recognize today."[51]

Some Christians may reject evolution because its process includes random events. But happenstance is inherent to authentic histories. Biological evolution is history, no less than that of manuscript traditions as revealed by textual critics, and in our histories "random seeking leads to non-random finding."[52] The gospel indicates that there is a destination to world history despite the freedom of its players. The trajectory heading to the purposed climax includes creation's evolving biota. In Christian terms, therefore, it is reasonable to consider self-aware, worshiping people as the goal of the phylogenetic process. Natural selection has led to our discovery of the personal dimension of reality.[53] We can gladly acknowledge phylogenetic history and its createdness. And the randomness that often characterises our own lives—accident, sickness, struggle—identifies those lives as authentic histories, sustained by their Creator, and destined for transformation in union with Christ.[54]

The author reports there are no competing interests to declare.
Received: 13/09/22 Accepted: 25/11/22 Published: 10/12/22

---

50 Andrew Briggs, Hans Halvorson, and Andrew Steane, *It Keeps Me Seeking* (Oxford University Press, 2018), 63, doi.org/10.1093/oso/9780198808282.001.0001.
51 Nick Spencer, *The Evolution of the West* (Louisville: Westminster John Knox, 2018), 8.
52 Andrew Steane, *Faithful to Science* (Oxford University Press, 2014), 71.
53 Briggs et al., *It Keeps Me Seeking*, 188.
54 1 Cor 2:7–9; Rom 5:3–5; 8:35–39; Jas 1:2–4, 12.

Imago Dei in the Age of Artificial Intelligence: Challenges and Opportunities for a Science-Engaged Theology

Marius Dorobantu

Abstract: Modern developments in evolutionary and cognitive science have increasingly challenged the view that humans are distinctive creatures. In theological anthropology, this view is germane to the doctrine of the image of God. To address these challenges, *imago Dei* theology has shifted from substantial toward functional and relational interpretations: the image of God is manifested in our divine mandate to rule the world, or in the unique personal relationships we have with God and with each other. If computers ever attain human-level Artificial Intelligence, such *imago Dei* interpretations could be seriously contested. This article reviews the recent shifts in theological anthropology and reflects theologically on the questions raised by the potential scenario of human-level AI. It argues that a positive outcome of this interdisciplinary dialogue is possible: theological anthropology has much to gain from engaging with AI. Comparing ourselves to intelligent machines, far from endangering our uniqueness, might instead lead to a better understanding of what makes humans genuinely distinctive and in the image of God.

Keywords: artificial intelligence; human distinctiveness; *imago Dei*; relationship; vulnerability

---

Marius Dorobantu is a Theology & Science researcher at the Vrije Universiteit Amsterdam and a fellow of the ISSR. His doctoral dissertation (at the University of Strasbourg) investigated the potential challenges of human-level AI for the theological understanding of human distinctiveness and the image of God. The article presented here was supported by the Templeton World Charity Foundation under Grant TWCF0542. The opinions expressed in this publication are those of the author and do not necessarily reflect the views of the Templeton World Charity Foundation.

In 2016, AlphaGo, a computer program developed by Google DeepMind, defeated one of the greatest human players of all time in the ancient strategy game of Go. For many, this event might not have been too significant. After all, computers had already mastered the much more popular game of chess for two decades, ever since Gary Kasparov's famous 1997 defeat by IBM's program, Deep Blue. For me, however, the news about AlphaGo was shattering. Having been an avid practitioner of the game for the best part of my life—both competitively and recreationally—I had a very good idea why this achievement was much more significant than Deep Blue's.

Originating more than four thousand years ago in China, the game of Go has deceptively simple rules. Two players, black and white, compete for limited resources by alternatively placing identical round pieces on a square board, trying to surround more territory than the opponent. Nevertheless, despite the simplicity of the rules, the ensuing complexity of the battle for territory dwarfs any other game. With each move, new possibilities open up, resulting in a cascading number of choices. There are more possible Go games than atoms in the observable universe.[1] For a long time, this made Go inaccessible to computers because the methods used to master other games, such as chess, were simply inapplicable to Go.

Traditionally, computers defeated human players in strategy games by leveraging their superior computing capabilities. Suppose a computer can go through all the relevant possible combinations of a situation on the board in a reasonable amount of time. In that case, there is no need for it to *understand* the game's principles or come up with clever strategies. It simply calculates all the possibilities and selects the one that most probably leads it to victory. In informatics terms, this is called *brute force*, and it is through brute force that Deep

---

[1] David Silver and Demis Hassabis, "AlphaGo: Mastering the Ancient Game of Go with Machine Learning," *Google AI Blog* (blog), 2016, http://ai.googleblog.com/2016/01/alphago-mastering-ancient-game-of-go.html.

Blue won against Kasparov.[2] In other words, a computer does not need to be clever if it can just laboriously explore all the possible routes. Due to its gargantuan complexity, Go does not lend itself to brute-force calculation. For this reason, the general feeling in the tech community was that it would take at least a few more decades until computers became capable of playing Go at a human level. Hence my surprise!

Besides the computational dimension, there was something more about AlphaGo's achievement that prompted the theologian in me to take notice, having to do with a more mystical aspect of the game of Go. When Go masters explain their moves, they rarely talk in mathematical terms. To be sure, their calculation abilities are outstanding and instrumental for success in the game. But Go masters often revert to a different kind of language when describing their play, one that belongs to the aesthetic register: *it felt good* to play there, or that move *looked beautiful*. A true Go master does not simply look to gain more points than the opponent; she looks for harmony on the board in a way not too different from a painter trying to achieve harmony on a canvas or a musician composing a masterpiece. Therefore, it is unsurprising that the game of Go was included among the four essential arts in ancient China, alongside music, calligraphy, and painting. There is as much calculation involved in a human master's game as intuition, creativity, and aesthetic taste. Moreover, there is arguably also a moral dimension to the game, at least when played by humans. A successful tactic presupposes an ideal mix of character virtues such as patience, humility, courage, and temperance. On the contrary, greed, arrogance, timidity, or pettiness are usually detrimental.

All the above are very subtle and elusive capacities that sit at the core of what we think it means to be human. It is hardly surprising that computers can beat us at chess by simply calculating the most relevant developments in advance. But if computers can beat us at Go, some hard questions arise about what they might become capable of in the

---

2   Paul Harmon, "AI Plays Games," *Forbes*, 2019, https://www.forbes.com/sites/cognitiveworld/2019/02/24/ai-plays-games/.

future and whether humans and computers are even that fundamentally different.

This article reflects on how progress in AI might impact the understanding of human distinctiveness in Christian theological anthropology, traditionally encapsulated in the notion that humans are created in the image of God (Latin, *imago Dei*). My central thesis is that theological discourse can benefit from engaging with the possibility of human-level AI, despite the apparent devastating impact such a scenario might exert on the idea of human distinctiveness. The analysis begins with a review of current *imago Dei* theology, demonstrating how theological discourse has hugely benefitted from engaging with evolutionary science. The following two sections reflect on how the two main modern interpretations of the divine image might deal with the emergence of intelligent robots. At this juncture, a question will be addressed: could AI be an equally good or even better image of God? The analysis concludes by stating that functional and relational *imago Dei* interpretations could still account for human distinctiveness from intelligent machines, but only insofar as they emphasise the importance of spiritual priesthood, authentic personal relationality, and vulnerability as fundamental human features, instead of rationality and intellectual prowess.

This conclusion demonstrates that theology can benefit from an honest engagement with AI and cognitive science, similarly to how it did by engaging with evolutionary science. Technological developments can bring beneficial limitations for theological speculation by rendering some hypotheses more plausible than others. In other words, it *is* possible for theologians to refine their understanding of human nature and distinctiveness by looking at the kind of intelligences that computer scientists are trying to build. This observation can strengthen the plea for a science-engaged theology. Furthermore, such conclusions regarding what it *really* means theologically to be human can constitute valuable contributions to the interdisciplinary debate on the future of technology. It is still unclear what truly constitutes the marker of humanness, or where does the threshold of personhood lie.

How we answer such questions as a global society will likely have significant ethical implications for how we treat each other, non-human animals, and robots. Theological anthropology can and should, therefore, bring its contribution to this all-important debate.

The Image of God after Darwin: Are We Still Special?

"What are human beings, that you are mindful of them?"[3] Since the age of the Psalmist, we have repeatedly asked this question with various methodologies: from theology and philosophy to biology, psychology, anthropology, and cognitive science. So far, none of these intellectual frameworks has come up with complete or satisfying answers.

From the perspective of evolutionary science, we are just one kind of living organism among many others, preoccupied, like all the others, with maximising its survival and procreation while inhabiting a rocky planet that orbits a typical star, just one of the hundreds of billions in the Milky Way. Biologically, we are essentially just another social ape with a slightly larger brain. What distinguishes us from all the other creatures is the things we can do, from writing poetry to sending people to the Moon or contemplating our death. However, all these impressive feats are made possible by anatomical structures and cognitive capacities we share with other creatures, even if they share those capacities in merely rudimentary form: nervous systems, language, mental representations etc. The point is that we do not seem to be as special as we thought we were.

This raises some problems for Christian anthropology because its central tenet is that humans *are* special. After all, they are created "in the image and likeness of God."[4] Since biblical times, we have had this intuition that there must be something special about us, something that distinguishes us from the rest of creation and makes us like our creator. The book of Genesis does not specify what exactly *imago Dei* is, but most interpreters thought of it in terms of some uniquely

3   Psalm 8:4.
4   Genesis 1:26.

human capacity having to do with our intellect, likely influenced by the Aristotelian tradition that regarded humans as rational animals.[5] This is known as the substantive interpretation of *imago Dei*. Nowadays, this interpretation has few adherents because most of the cognitive capacities thought uniquely human in the prescientific age have recently been fully or partially identified in other animals. Furthermore, since Darwin, it has become clear that humans are not ontologically different from the rest of living creatures. We are part of the same tree of life and share most of our DNA—up to 99%—with non-human species.[6]

What does it mean then to be in the image of God, if not to possess some exceptional intellectual faculty? To replace the problematic substantive interpretation, theologians have creatively devised more sophisticated accounts of human distinctiveness and *imago Dei*, most of which broadly fall into two big categories: functional and relational. The functional interpretation locates our specialness not in our mental capacity, but in our election by God,[7] and in what we are called to do, namely, to represent God in the world by exercising dominion and stewardship over the rest of creation. This idea is rooted in the modern biblical exegesis of the notion of image. The assumption is that the image in Genesis was used with a meaning inspired from other cultures in the ancient Near East. To be the image of a particular god, typical of kings or pharaohs, was to represent that god on earth and exercise authority on that god's behalf.[8]

---

5   For reviews of *imago Dei* interpretations, see Noreen L Herzfeld, *In Our Image: Artificial Intelligence and the Human Spirit* (Minneapolis: Fortress Press, 2002); Marc Cortez, *Theological Anthropology: A Guide for the Perplexed* (A&C Black, 2010); J. Wentzel Van Huyssteen, *Alone in the World? Human Uniqueness in Science and Theology* (Grand Rapids, MI and Cambridge: William B. Eerdmans, 2006); Stanley J Grenz, *The Social God and the Relational Self: A Trinitarian Theology of the Imago Dei* (Louisville, KY: Westminster John Knox Press, 2001).

6   Robert H. Waterson et al., "Initial Sequence of the Chimpanzee Genome and Comparison with the Human Genome," *Nature* 437:7055 (2005): 69–87, https://doi.org/10.1038/nature04072.

7   Joshua M. Moritz, "Evolution, the End of Human Uniqueness, and the Election of the Imago Dei," *Theology and Science* 9:3 (2011): 307–339, https://doi.org/10.1080/14746700.2011.587665.

8   Claus Westermann, *Genesis: An Introduction* (Fortress Press, 1992), 36–37; David J. A. Clines, "The Image of God in Man," *Tyndale Bulletin* 19 (1968): 93.

The other option, the relational interpretation, regards the image of God as manifested in the unique relationship humans are called to have with God and in the authentic personal relationships they have with each other.[9] God, the Holy Trinity, is relationship, and so is humanity because "in the image of God he created them, male and female he created them."[10]

Both these interpretations of *imago Dei* provide better answers to the scientific challenges mentioned earlier than the substantive interpretation. Human distinctiveness does not reside in any uniquely human intellectual faculty but in our unparalleled agency in the world, which we are called to care for and even co-create with God (functional interpretation), or in the relationality that is so central to what it means to be human, and in which we mirror a Trinitarian God (relational interpretation). Although, indeed, we are not the only species that significantly acts upon its environment—many animals, for example, engage in what is known as niche-construction[11]—the sheer scale of our dominion over the earth, at least since the agricultural revolution onwards, might be seen as a proof of our special vocation. Similarly, although we are not the only relational species, the complexity of our personal relationships and the importance of relationships in the development and flourishing of the human person seem to support the idea that it is through our relationality that we are special and in the image of God.

The functional and relational interpretations of the image arguably represent progress from the earlier substantive proposal. This shows that theological anthropology ultimately stands to gain from an open and honest engagement with science. As English theologian Aubrey Moore aptly put it more than a century ago, "Darwinism appeared, and, under the guise of a foe, did the work of a friend."[12] Rev-

---

9   Karl Barth, *Church Dogmatics*, vol. 3 (Edinburgh: T&T Clark, 1958).
10   Genesis 1:26.
11   Michael Burdett, "Niche Construction and the Functional Model of the Image of God," *Philosophy, Theology and the Sciences (PTSc)* 7:2 (2020): 158–180, https://doi.org/10.1628/ptsc-2020-0015.
12   Francisco J. Ayala, *Darwin's Gift to Science and Religion* (Joseph Henry Press, 2007), 159.

olutionary scientific ideas, such as Copernicus' heliocentric theory or Darwin's evolutionary theory, may appear at first to menace long-held religious beliefs about the world and the human being. Still, once the dust settles, theological reflection is actually enriched by the process of incorporating new scientific knowledge. As it turns out, it is still perfectly possible to speak of a creator God even when we know the cosmos is much older than a few thousand years. Likewise, there are new and arguably better theological ways of speaking of human distinctiveness, even when evolutionary theory shows that we are of the same ilk as nonhuman creatures, and that our cognitive abilities are not that much different in kind from theirs.

However, a new type of challenge for human distinctiveness looms large on the horizon, as hinted at earlier in the AlphaGo story. Starting with the 1950s, computer programs have become capable of matching and surpassing human abilities in an increasing range of tasks, which, when done by humans, require what we vaguely call intelligence. We call this type of program Artificial Intelligence (AI). Even if AI operates somewhat differently from biological intelligence, AI programs are astonishingly capable of doing many of the things we used to regard as the unique domain of human intelligence, such as solving problems, proving theorems, labelling the content of images, transforming speech into text, translating various languages, composing music, and answering questions, to name just a few.

If progress in AI continues, it is not entirely absurd to imagine a time in the future when computers will reach human-level intelligence, becoming able to do all the things that we do equally well or even better. To a certain extent, this is already happening in some domains. AI algorithms can diagnose some forms of cancer better than human doctors.[13] They operate at a superhuman level in chess, Go, and many other strategy games. We trust AI programs to land planes and run the stock markets because of their ability to make fast decisions

---

13 Scott Mayer McKinney et al., "International Evaluation of an AI System for Breast Cancer Screening," *Nature* 577:7788 (2020): 89–94, https://doi.org/10.1038/s41586-019-1799-6.

better than error-prone humans. One day, our streets might be filled with the much-hyped autonomous cars, or we might engage in deep spiritual conversations with our robotic companions.

When thinking about the challenges posed by AI to the idea of human distinctiveness, the hypothetical scenario of human-level AI is undoubtedly of great relevance. Nonetheless, an argument can be made more broadly that even without such spectacular developments, AI is still relevant for theological anthropology. Here, I would like to refer to AI as more than just the intelligent machines themselves. Instead, AI designates the fundamental study of the nature of intelligence performed by trying to endow machines with intelligence. This is precisely how the field of AI set off in the 1950s. Alan Turing, one of the founders of theoretical computer science and AI, believed that trying to create a thinking machine could shed light on how humans think.[14] In this respect, AI can be seen as an applied form of cognitive science,[15] and its results can be interpreted as saying something relevant about how humans achieve cognition. If AI easily masters chess, Go, prose, or visual arts, this can produce meaningful clues about the nature of such endeavours. On the contrary, if AI stumbles at particular tasks, that is also relevant, perhaps pointing to features that pertain to human distinctiveness. Therefore, both through its successes *and* failures, AI can produce new data points, which can further serve as food for insightful theological reflection.

Could Robots Be Better Images of God?

If AI does reach human level performance, that is, if it matches our ability to *do* things, then the functional interpretation of the image of God as human distinctiveness may become problematic. As long as

---

14 Jack Copeland, *Artificial Intelligence: A Philosophical Introduction* (Blackwell, 1993), 26.
15 Trying to endow computers with intelligence is one approach. Another approach is the attempt to simulate on supercomputers the neural connections in the mammalian brain: Nidhi Subbaraman, "Artificial Connections," *Communications of the ACM* 56:4 (2013): 15–17, https://doi.org/10.1145/2436256.2436261.

we remain the most capable creature on earth in terms of the things we do, we can still see this as marking our distinctiveness and kinship with God. But how about a scenario where we became stripped of this privileged position by our artificial "mind children"?[16] What if robots became better than humans at ruling the world and, thus, better representatives of God? Should they not, then, also be considered in the image of God (perhaps even more than us?) according to the functional interpretation?

The above hypothesis might look like the stuff of sci-fi movies, but many people in AI take it seriously. In a 2014 survey, 550 AI experts were asked to predict the likelihood of AI reaching the human level soon. The 2040s got a 50% median probability, while the year 2075 got a 90% probability.[17] There is no way of knowing how AI development will continue. Maybe it will slow down and plateau, never really getting anywhere close to the human level. But there is also the opposite scenario, known as the "intelligence explosion,"[18] where progress in AI accelerates, maybe due to machines becoming better than humans at programming AI, thus triggering a positive feedback loop of self-improvement. This scenario is also referred to as the technological "singularity."[19] According to philosopher Nick Bostrom, there is a real possibility that AI could reach a super-human level sometime in the future, something he calls artificial super-intelligence (ASI).[20] We, humans, are severely limited regarding how intelligent we can become. The amount of knowledge we can acquire in a lifetime is limited, our brains cannot grow bigger than our skulls, and they inevitably decay and die after

16 Hans Moravec, *Mind Children: The Future of Robot and Human Intelligence* (Harvard University Press, 1988).
17 Vincent C. Müller and Nick Bostrom, "Future Progress in Artificial Intelligence: A Survey of Expert Opinion," in *Fundamental Issues of Artificial Intelligence*, ed. Vincent C. Müller (Cham: Springer International Publishing, 2016), 555–572, https://doi.org/10.1007/978-3-319-26485-1_33.
18 Ronald Cole-Turner, "The Singularity and the Rapture: Transhumanist and Popular Christian Views of the Future," *Zygon* 47:4 (2012): 787, https://doi.org/10.1111/j.1467-9744.2012.01293.x.
19 Ray Kurzweil, *The Singularity Is Near: When Humans Transcend Biology* (Penguin, 2005).
20 Nick Bostrom, *Superintelligence: Paths, Dangers, Strategies* (Oxford University Press, 2014).

several decades. Machines do not share such limitations, and so, in principle, ASI could become more intelligent than any human being, than all of humanity collectively, and even intelligent beyond human comprehension. Bostrom demonstrates quite convincingly that any attempt on our part to contain and control ASI would ultimately be futile because such a super-intelligent agent could see straight through our plans and anticipate any potential strategy we might devise.

There are legitimate concerns about the existential risk posed to our species by ASI, but there are also formidable things that ASI could do for us. The ascension of artificial minds may not happen through a violent rebellion, as often depicted in futuristic movies, but rather with our blessing and cooperation. As our world becomes more complex and data-driven, we will rely increasingly on artificial systems to assist us in our decisions or even to make them in our stead. I mentioned earlier the example of stock markets, which are run by such AI programs, but many other aspects of our lives are already governed mainly by algorithms: what we see in our social media feeds, the music and movies recommended to us by streaming services, how much money we can borrow from a bank, or which medical procedure to choose based on our profile. We are becoming increasingly aware of all the ethical problems associated with this, but it does not seem that we have any intention to reverse this trend anytime soon. Although the loss of privacy and decision-power bothers us in principle, the convenience facilitated by these apps is often too appealing. This is precisely why it is not hard to imagine a future when most, if not all, power is voluntarily granted to AI systems, especially if their competence keeps improving.

Bostrom speaks of three ways ASI might operate: as an oracle, a genie, and a sovereign. As an oracle, it would answer all our questions; as a genie, it would execute all our commands; as a sovereign, it would govern the world with "an open-ended mandate to operate ... in pursuit of broad and possibly very long-range objectives."[21] Those with a trained theological eye might notice an eerie resemblance to the kind of role ascribed to God in monotheistic religions. But leaving the issue

---

21   Bostrom, *Superintelligence*, 181.

of idolatry aside, the possibility of ASI governing the world better than we do seems deeply problematic for the functional interpretation of *imago Dei*. How could we still claim to be exceptional if AI proves to be a better steward of creation?

The task is not even that hard to fulfil, given how disastrously we have been performing so far. In our exploitation of animals, we have caused tremendous suffering, especially in the last few decades, with industrial farming. In our greed, we are currently driving the atmosphere to heat up, endangering the ecological balance on a global scale. These *achievements* are hardly something worthy of the divine mandate to represent God in the world. ASI could do a better job, at least in theory. And while that might be something to hope for, from a theological perspective it raises some hard questions about human distinctiveness and our role as stewards of creation appointed through divine election. How could we still speak of such things in a scenario of more-competent-than-humans AI?

I think the question is legitimate, but I do not think a scenario of human-level AI completely invalidates a functional understanding of the image of God. The reason has to do with the scope of our divine mandate to rule over the world, at least as it is understood in many Christian traditions. Our vocation to care for creation goes beyond the historical realm and ultimately has a spiritual dimension. The Romanian-Orthodox theologian Dumitru Stăniloae speaks of a priestly vocation that we are called to, one that enables and compels us to raise the world to a "supreme level of spiritualisation":

> The world was created in order that humanity, with the aid of the supreme spirit, might raise the world up to a supreme spiritualisation, and this to the end that human beings might encounter God within a world that had become fully spiritualised through their own union with God. The world is created as a field where, through the world, humankind's free work can meet God's free work with a view to the ultimate and total encounter that will come about between them. For if humanity were the only free agent working within the world, it could not lead the world to a

complete spiritualisation, that is, to its own full encounter with God through the world. God makes use of humanity's free work within the world in order to help humanity, so that through humanity's free work both it and the world may be raised up to God and so that, in cooperation with humankind, God may lead the world toward that state wherein it serves as a means of perfect transparency between humanity and himself.[22]

Humans are not called to simply govern and organise creation in a worldly fashion. Instead, they are given the higher task of uplifting creation to complete spiritualisation. There is a remarkable convergence between this kind of theological language and the language used by some of the most prominent prophets of AI and the singularity. Futurists like Ray Kurzweil[23] or James Lovelock,[24] for example, believe that the cosmos longs for informatisation and that only future cyborgs or robots will be capable of saturating the universe with intelligence. Humanity's role, in their view, is that of a midwife to superior, synthetic forms of intelligence that will expand to corners of the universe inaccessible to biological life. Is this informatisation of matter the same as the spiritualisation invoked in Christian theology? I think not.[25]

Firstly, information does not equal spirit, despite both pointing to something immaterial. Nowadays, there exists this tendency to believe that anything that transcends the material domain must be informational. For example, the soul or the mind is sometimes regarded simply as informational pattern, which explains why some people in the transhumanist movement believe their minds could be uploaded to a computer. The Christian notion of spirit is much richer than the

---

22 Dumitru Stăniloae, *The Experience of God: The World: Creation and Deification* (Holy Cross Orthodox Press, 2000), 59 (slightly altered).
23 Kurzweil, *The Singularity Is Near*, 21.
24 James Lovelock, *Novacene: The Coming Age of Hyperintelligence* (Penguin UK, 2019).
25 I argue this in more detail in Marius Dorobantu, "Why the Future Might Actually Need Us: A Theological Critique of the 'Humanity-As-Midwife-For-Artificial-Superintelligence' Proposal," *International Journal of Interactive Multimedia and Artificial Intelligence* 7:1 (2021): 44-51, https://doi.org/10.9781/ijimai.2021.07.005.

idea of information, pointing to a transcendent dimension of reality. Secondly, as evident in Stăniloae's account, Christian theology embeds the spiritualisation of matter in the love relationship between God and humans. Spiritualising the world is not an end in itself, but rather a means to achieve complete transparency between creator and creation. Without God's love and purpose for creation, any spiritualisation/informatisation of matter is empty of content and significance. What would be the finality of such a process? A state of perfect and eternal cosmic equilibrium? In physics, such a scenario is known as the "big freeze," and it is synonymous with a heat death of the universe, where nothing more can happen due to a state of maximum entropy.[26] How could this be a cosmic state we should be rushing towards?

The theological account of the mystical role of humans in the world seems thus much more cogent than its secular counterparts. For theological anthropology, the implication is that a functional interpretation of the image of God needs to focus more on the spiritual dimension of our dominion and stewardship and not so much on its historical side, where AI may indeed outmatch us. The other dimension that needs to be stressed more concerning our vocation is the relational one. Our role in creation should not be divorced from our relationship with God. Being in the image of God does not entail just having been elected as God's representative at a certain point in or outside history. Instead, as shown by Stăniloae, it involves a continuous personal, authentic relationship of love between creature and creator, which brings us to the relational interpretation.

Vulnerable God, Vulnerable Humans, and the Image as Relationship

In a relational interpretation, the divine image is to be found in the loving relationships we develop with God and each other. Profound relationality is the mark of human life. We are born as a result of re-

---

26 A. V. Yurov, A. V. Astashenok, and P. F. González-Díaz, "Astronomical Bounds on a Future Big Freeze Singularity," *Gravitation and Cosmology* 14:3 (2008): 205–212, https://doi.org/10.1134/S0202289308030018.

lationships, our personhood can only develop in relationships, and it is mostly in our relationships that we find meaning, purpose, and fulfilment. If it is through relationships that we best mirror God, then developments in AI might legitimately question our distinctiveness. What if machines become one day capable of personal relationships? We already have conversations with chatbots, and the complexity of these conversations only increases as technology gets better. It is not unimaginable that in the future, we might talk to machines as we currently talk to humans.

This is precisely what Alan Turing proposed as a litmus test for whether a machine is truly intelligent. If someone conversing via text with the AI cannot tell whether they are talking to a human or a machine, then that machine should be considered intelligent.[27] This has become known as Turing's test and is still widely regarded as a valid benchmark for human-level AI. As of today, no program has passed the test, but as shown earlier, many people believe it to be just a matter of time before it happens. Would an AI capable of human-level conversations really engage in personal, authentic relationships? This is a tricky question, as shown by the confusion and heated debate that recently ensued when a Google engineer publicly expressed his concern that LaMDA, an AI he was working with, had become sentient.[28]

On the one hand, there are good reasons to believe that just displaying relational-like behaviour does not mean that an authentic relationship is actually being formed. Intuitively, a genuine self or consciousness is needed for the I-Thou type of relationship. Humans are such selves, while inanimate objects are not. Humans are *someone*, while machines are *something*. In Ted Peters' words, "nobody is at home" inside these machines.[29] On the other hand, we lack a convincing scientific theory to explain this difference between the presence

27 Alan M. Turing, "Computing Machinery and Intelligence," *Mind*, stb, 59:236 (1950): 433–60, https://doi.org/10.1093/mind/LIX.236.433.
28 Nitasha Tiku, "The Google Engineer Who Thinks the Company's AI Has Come to Life," *Washington Post*, November 6, 2022, https://www.washingtonpost.com/technology/2022/06/11/google-ai-lamda-blake-lemoine/.
29 Ted Peters, "Will Superintelligence Lead to Spiritual Enhancement?" *Religions* 13:5 (2022): 5, https://doi.org/10.3390/rel13050399.

and absence of consciousness, phenomenal experience, or subjectivity. In other words, we do not really know what makes us persons and conscious agents. What is the secret ingredient that we possess and that robots lack? In the philosophy of mind, this is famously known as the "hard problem of consciousness,"[30] namely, how can consciousness or subjective experience arise from inert matter? Theologically, this problem can sometimes be dismissed with more ease if we believe in the existence of an immaterial soul. A supernatural soul could be a convenient explanation for the hard problem of consciousness. But unless one commits to a strong form of mind-body dualism that is at odds with most contemporary philosophy, speaking of a soul is plagued by the same kind of questions. Until we have a clearer understanding of what constitutes an authentic self, it is not wise to pontificate that machines will never become such selves.

People often point to the fact that AI is purely algorithmic and deterministic, thus incapable of consciousness, personhood, or freedom. But the same argument can be turned against humans because, from a scientific/mechanistic perspective, we are also algorithmic and, to some extent, deterministic beings, with the only difference that our algorithms are biological, genetic, or neurological, rather than digital or electronic. I do not necessarily subscribe to this view, but it is indeed tough to argue against it on purely scientific grounds. Insofar as the natural sciences are concerned, both humans and computers are machines, just different types. One needs to look at the issue from a completely different vantage point, for example of theology, to see something truly special about human beings. For the reasons listed above, it would be tough to decide whether an AI that acted as if it were conscious really was, or whether it was just simulating it. A robot claiming to be in love, suffer, or believe in God would pose challenging ethical, philosophical, and theological problems.

I think that contrary to what sci-fi likes to depict, the above scenario is improbable. AI is currently developing to *think* very differently

---

30 David Chalmers, "Facing Up to the Problem of Consciousness," *Journal of Consciousness Studies* 2:3 (1995): 200–219.

from how humans think. When labelling images, playing Go, or responding to text messages, a human and a computer program might sometimes produce the same result, but with very different tools and methodologies. Even when AI manages to attain human-level competency in various domains, it does so in a very non-humanlike fashion. When it makes mistakes, those are not the kind of mistakes that any human would make. Even if we somehow managed to endow our artificial creatures with a self and phenomenal experience, those would likely be radically different from our own due to our very different types of embodiment. Robots would have different perceptual senses, a different kind of access to their memories or internal states, and a very different relationship with time. Their needs would differ from ours, significantly impacting their interests and motivations. AI might indeed reach human-level competency someday, but it will probably be very non-humanlike.[31]

This is good news for the relational interpretation of *imago Dei* because it means that the kind of personal relationships that we have with each other will not necessarily be part of the robots' behavioural repertoire. Our relationality is very much connected with our vulnerability. We engage in relationships precisely because we are vulnerable and mortal, and need one another. There can be no genuine relationship without the two partners making themselves vulnerable to each other beyond any transactional logic. This is why deep relationships are always risky, because of the looming possibility of getting hurt. But without such voluntary vulnerability, how could anything deep and meaningful ever emerge? How could an artificial being, which is virtually invulnerable and immortal—having copied backups of itself on multiple computers—engage in humanlike relationships?

In Christian theology, this powerful idea that vulnerability is instrumental for authentic relationality is manifest in the doctrine of the incarnation. God does not shy away from vulnerability, but quite the

---

[31] Marius Dorobantu, "Human-Level, but Non-Humanlike: Artificial Intelligence and a Multi-Level Relational Interpretation of the Imago Dei," *Philosophy, Theology and the Sciences (PTSc)* 8:1 (2021): 81–107, https://doi.org/10.1628/ptsc-2021-0006.

contrary. Through Jesus Christ, we see God subjecting Godself to the ultimate vulnerability of suffering and death on the cross out of love for creation. As humans, we image God when we are loving and vulnerable, not when we are mighty and unbreakable.

Besides vulnerability, another reason why human-level AI will likely not share in the kind of personal, humanlike relationships is its hyper-rationality. It is unlikely that a creature who makes all its decisions based on cold calculations of optimal outcomes will engage in such risky and irrational behaviour. We humans seek relationships because we have a sense of incompleteness and deep hunger for a kind of fulfilment that cannot be achieved solely within ourselves. Unlike the AI, we do not entirely understand our internal states and motivations, so we try to know ourselves better in relationships with others. That incompleteness drives us to seek the companionship of other humans, and it is arguably one of the main drivers of our religiosity, of why we seek God. This restlessness of our hearts, as Augustine called it,[32] or what Wolfhart Pannenberg refers to as exocentricity,[33] comes from deep within ourselves, from way below our rational minds. A purely rational being would not behave like this. Falling in love is certainly not a rational thing to do. However, it is such irrational things, from love to art to spirituality, that make human life enjoyable. Perhaps it is precisely because we are *not* as intelligent as AI that we can image God relationally.

The exciting developments in the field of AI arguably represent a blessing in disguise for theological anthropology, and this also constitutes an opportunity for a science-engaged theology. Far from endangering human distinctiveness, AI helps us appreciate some of the things that make us human and, therefore, different from machines. Following Aristotle, many Church Fathers believed that it is through our rationality and reason that we image God because that is what dis-

---

[32] "You have made us for yourself, and our heart is restless until it rests in you." Saint Augustine, *Confessions* (Oxford University Press, 2008), 3.
[33] Wolfhart Pannenberg, *Anthropology in Theological Perspective* (Westminster Press, 1985), 51.

tinguishes us from the animals.[34] What reflection on AI shows is that, although we might be more rational than nonhuman animals, we are certainly not the apex of rationality. Furthermore, because we are not entirely rational, we can engage in authentic relationships with other human persons and with God. In doing this, we mirror God, our creator, and become and flourish as authentic persons. Humans might look irrational and outdated when compared to the AI. Still, it is precisely because of our bodily and cognitive limitations that we can cultivate our divine likeness through loving, authentic, personal relationships. If reflecting on AI teaches theologians one thing, it is that our limitations are just as important as our abilities.[35] We may be vulnerable, but in being so we resemble a vulnerable God.

In my opinion, the truly ground-breaking conclusion from reflecting theologically on AI is that being like God does not necessarily mean being more intelligent. Christ's life and teaching show that what is most valuable about human nature are traits like empathy, forgiveness, and meekness, which are all eminently relational qualities. What enables such attributes is a kind of thinking rooted more in the irrational than in the rational parts of our minds. Perhaps this can shed new light on Saint Paul's choice to "boast all the more gladly of my weaknesses ... or whenever I am weak, then I am strong."[36]

Conclusion

Although AI does not, in principle, challenge our theological understanding of human distinctiveness, our attitude towards this technology raises an important alarm about the future of human self-reflection. We are very much in awe of these machines and ready to consider them intelligent only until we understand how they work. In this sense, true AI has been an ever-receding horizon so far because our standards of

34 For example, Saint Gregory of Nyssa, *On the Making of Man* (CreateSpace Independent Publishing Platform, 2013).
35 For a detailed argumentation, see Marius Dorobantu, "Cognitive Vulnerability, Artificial Intelligence, and the Image of God in Humans," *Journal of Disability & Religion* 25:1 (2021): 27–40, https://doi.org/10.1080/23312521.2020.1867025.
36 2 Corinthians 12: 9–10.

what indeed constitutes intelligence are continuously shifting. John McCarthy, who coined the term *artificial intelligence*, says it best: "as soon as it works, no one calls it AI anymore."[37] If we could travel back in time and show people fifty years ago the iPhone chatbot Siri, they would surely be astonished and consider it true AI. But for us, today, it is just another app. This is because we have looked behind the curtain, and we know more or less how it works: there is no magic involved! The more we understand how something works, the less inclined we are to ascribe it intelligence and value. This tendency is worrying because sometime in the future, it might be humans, instead of machines, that we disregard.

Our world is built around humanistic values, which stem from our fascination for the ultimate mystery of the human being. There are still so many things that we do not understand about ourselves, especially regarding our minds: what is the nature of thoughts, how are memories stored, how do we make decisions etc.? Human beings escape any complete theory or explanation, and this persisting mystery is probably one of the main reasons why we grant dignity and rights to human persons. Neuroscience and psychology are still in their infancy, but what if someday we *did* acquire a complete knowledge of the human person? What if we realised that we were, in fact, automata obeying algorithms that, although unspeakably complicated, are still ultimately deterministic? Should we then do away with human dignity and rights and treat humans as we currently treat other creatures and objects that we consider mindless? Obviously not. And this is precisely why theological anthropology should insist on an understanding of human distinctiveness and *imago Dei* rooted not in what humans are *on the inside*, as in the structural interpretation, but in our special relationship with God and the value of our relationships with each other. A move towards such a relational ontology would not only disentangle

---

[37] Moshe Y. Vardi, "Artificial Intelligence: Past and Future," *Communications of the ACM* 55:1 (2012): 5, https://doi.org/10.1145/2063176.2063177.

human dignity from human intellectual exceptionalism, but it would also arguably be more faithful to Christian Trinitarian theology.[38]

Lastly, there is one area, in particular, where theological anthropology could bring a valuable contribution to the global discussion of our future with AI.[39] As Bostrom and many others have warned, there is a real danger in granting too much power to a technology whose control we could quickly lose. The worry is not that robots will consciously rebel against us like in the movies, but more that they might harm us involuntarily while trying to do exactly what we asked them to do. Concepts and values that would be obvious to a human being are not necessarily evident to a computer. That is why many brilliant computer scientists and philosophers are currently working on the so-called *AI alignment* problem. They try to ensure that even if machines eventually escape our direct control, their values will be sufficiently aligned with our own that they will not accidentally harm either us or anything else important to us. However, when it comes to which exact values to bake into these algorithms, things become complicated very quickly because there is no universal set of human values shared across cultures. It goes without saying that religious traditions should be part of this conversation because of the many people they represent and their centuries of experience reflecting on human values.

With all the noise generated by realising the potential threats of artificial super-intelligence, a more subtle danger goes completely unnoticed. Because most attention is devoted to preventing a catastrophic scenario, a consensus seems to uncritically emerge that an ASI that did not kill us would necessarily be good. We seem to be so caught into the otherwise crucial problem of aligning AI to our goals that we often do not even question whether we should even attempt to build ASI in the

---

38 John D. Zizioulas, *Communion and Otherness: Further Studies in Personhood and the Church* (Bloomsbury Publishing, 2010).

39 For a broad discussion of issues in AI and Christian theology, see Marius Dorobantu, "AI and Christianity: Friends or Foes," in *Cambridge Companion to Religion and AI*, ed. Beth Singler and Fraser Watts (New York: Cambridge University Press, forthcoming); Marius Dorobantu, "Artificial Intelligence as a Testing Ground for Key Theological Questions," *Zygon* 57:4 (2022): 984–999, https://doi.org/10.1111/zygo.12831.

first place. The assumption is that it is good to bring about Bostrom's oracle/genie/sovereign because of the age of abundance, peace, and leisure that would follow. ASI would govern and feed us, take care of our energy needs, and in general solve all the complex problems in our stead so that we could devote our lives to more pleasant endeavours. We would effectively be ASI's pets.[40] Who could possibly argue against such a future? How could the eradication of poverty and sickness not be a good thing? Although it is difficult to deny a certain appeal to this idea, many people would feel that something is just not right with this kind of *brave new world*. But this intuition cannot be articulated without an appeal to questions about what a good life is, the purpose of human existence, the value of vulnerability and suffering, and why freedom is ultimately more precious than comfort. To me, these are all theological questions and represent an exciting entry point for theology into the interdisciplinary and global dialogue on new technologies.

The author reports there are no competing interests to declare.
Received: 25/08/22 Accepted: 30/11/22 Published: 12/12/22

---

[40] Samuel Gibbs, "Apple Co-Founder Steve Wozniak Says Humans Will Be Robots' Pets," *The Guardian*, June 25, 2015, sec. Technology, https://www.theguardian.com/technology/2015/jun/25/apple-co-founder-steve-wozniak-says-humans-will-be-robots-pets.

Book Reviews

Jacob Shatzer: *Transhumanism and the Image of God*

Downers Grove: IVP Academic, 2019; 192 pages.
ISBN-13 0830852505.

Many people would see Transhumanism, if they've heard about it at all, as a fringe movement populated by weird technophiles and people (like me) who've never outgrown their adolescent passion for science fiction. Even those with a more than passing knowledge of it generally see it as an extreme, almost cult-like phenomenon that has little real-world impact, and few, if any, serious implications for Christian discipleship. In this book Jacob Shatzer (Assistant Professor and Associate Dean in the School of Theology and Missions at Union University) seeks to introduce a general Christian audience to the world of transhumanism and, perhaps more to the point, show how current technologies and their likely near-future developments reflect transhumanist ideals and make the attaining of those ideals more plausible (and possible).

The book is well-constructed and clearly arranged, and makes a helpful contribution to critical engagement with technology.

The introduction does a fine job of doing what an introduction ought to do: it engages our attention and outlines the argument of the book. Chapter 1, *Technology and Moral Formation*, picks up where the introduction leaves off, presenting a more substantial account of how technologies and the forms of life associated with them shape us as moral agents. Eight substantive chapters follow, falling into two main sections. The first section deals with transhumanism: 2 *What Is Transhumanism?*; 3 *My Body, My Choice: Morphological Freedom*; 4 *The Hybronaut: Understanding Augmented Reality*; 5 *Meeting Your (Mind) Clone: Artificial Intelligence and Mind Uploading*. The second section addresses current technologies and how we should critically engage with (and disengage from) them for the sake of discipleship: 6 *What Is Real? Changing Notions of Experience*; 7 *Where Is Real? Changing Notions of Place*; 8 *Who Is Real? Changing Notions of Relationships*; 9 *Am I Real? Changing Notions of the Self*. The book closes with a *Conclusion: The Table*, in which the author reflects on particular practices that he believes will help us faithfully navigate the technological and cultural landscape that lies before us.

Space precludes even a summary of his key contentions, let alone an analysis, so I will confine myself to identifying a few important contributions Shatzer makes, as well as pointing out some weaknesses and limitations in the book.

It is important to keep Shatzer's primary goals in mind: critically introducing the key tenets of transhumanism, and demonstrating the ways we are being shaped in line with transhumanist goals (and suggesting ways we might become alert to that [mal-] formation and enact strategies and practices that might resist it). He does not intend to give an exhaustive introduction to transhumanism, nor a comprehensive theological account and critique of it. Nor does he aim to present a social and theological appraisal of current and emerging technologies. Rather, he seeks to equip Christians to engage faithfully with those technologies in light of what he sees as their fostering of transhumanist ideals and their impetus towards transhumanist goals. His aim is, in short, to foster critical discernment for the sake of Christian discipleship, a goal he largely achieves.

Shatzer's opening reflection on time and the history of timepieces, while seemingly unrelated to transhumanism, is an effective introduction to key themes that underlie any sound analysis of technology and human society: it helps us see the way that technology is far from neutral: while a product of humans and human culture, it shapes both our culture and us as persons. Using something as ubiquitous and commonplace and seemingly trivial as clocks provides a good introduction to that key contention, which, when brought to bear on transhumanism and digital technologies, leads well into the central thesis of the book: "I argue that Christians must engage today's technology creatively and critically in order to counter the ways these technologies tend toward a transhuman future." While there are aspects of what follows that are open to question, this is a largely helpful introduction to the issues.

Of equal value is the way that autonomy and the desire for unfettered control permeate his discussion of both transhumanism and the technologies he addresses. It is easy for us to be blind to the pervasive influence of these deeply embedded features of 21st century techno-capitalism and western-influenced cultures. Shatzer's consistent exposure of their presence serves to (re) awaken our critical faculties and so, please God, help us resist their malforming effects.

It is not, however, a flawless book. His use of concepts and terminology can be idiosyncratic (as, for instance, in his understanding of transhumanism and its relationship with posthumanism), and some of his claims, and the arguments that support them, need more work. Moreover, his is a largely negative assessment of the technologies he assesses. While he makes some good points, I would have liked to have seen clearer discussion of the ways that these technologies can be "disciplined" towards ends that foster human flourishing and might contribute to Christian discipleship. For instance, online multi-player gaming platforms and the virtual "communities" that form around them can and have been used as vehicles for discipleship. And, since the advent of Covid, the possibilities of live-streaming and online interaction have become apparent. To be fair, the book was published in 2019, and much has changed in the life of the church in those few years, but some of the possibilities antedate the pandemic.

Perhaps more importantly, critical engagement with technology with a view to Christian discipleship needs to do more than *criticise* and suggest disciplined patterns of abstinence (useful as these may be). It needs to recognise potential and, as Andy Crouch has persuasively argued, engage in positive forms of culture-making so as to create cultural products that positively form people and their communities (see *Culture Making*, Downers Grove: IVP, 2008). While one book can only do so much, I would have liked to have seen some clearer nods in these directions.

It is also puzzlingly less well-integrated than it should be. There are a number of threads left dangling in individual chapters that could—and should—have been woven through other chapters and, in some cases, through the book as a whole. For instance, while Shatzer notes the ways that market forces are in play in social media contexts, he does not bring that to bear on other technologies, such as body-morphing. Perhaps of more significance is his failure to bring his discussion of the profound embodiment of human cognition to bear on his treatment of uploaded consciousness. And there are other instances. These, along with simple repetition of key quotes on a number of discrete occasions gives the book a piecemeal feel. An obvious point of integration would be to bring theological anthropology into the discussion. There are passing references to embodiment, the Incarnation, the Fall, and the like (e.g., in Ch. 6), but no sustained treatment of the topic—which is surprising given the title of the book: *Transhumanism and the Image of God*. Perhaps discipleship was meant to serve that function (if so, should it have been in the title?); but his discussion of discipleship is largely topical rather than integrative, lending to the piecemeal feel of the book.

Even so, this is a useful book and worth reading. Those who are familiar with transhumanism and/or the critical analysis of technology may learn something; they may also find a provocative conversation partner. And those who aren't, will find a useful guide to the strange world inhabited by technophiles and us post-adolescent sci-fi junkies.

Andrew Sloane
Morling College
January 2022

Ronald L. Numbers (Ed.): Galileo Goes to Jail: And Other Myths about Science and Religion

Cambridge: Harvard University Press, 2010; 298 pages.
ISBN-13: 9780674057418.

In *Galileo Goes to Jail*, Numbers includes essays from the breadth of scholarship in the history of science and religion. This book is one that no one with an interest in science and religion should go past.

The book consists of 25 essays about various myths contributing to widespread misunderstanding of the relationship between theology and science. Numbers' six-page introduction is one of the best brief summaries of the field I have ever read. Everyone should read it.

Normally a collection of essays will have uneven scholarship. This is not the case with this book. Each essay is well written by an expert on that aspect of the field. The contributors include David Lindberg, Margret Osler, Lawrence Principe, Peter Harrison, Nicolaas Rupke, James Moore, Michael Ruse and John Hedley Brooke. Anyone reading in the field should read more by all these contributors. *Galileo Goes to Jail* is a great introduction to the breadth of scholarship in science and religion.

A very broad selection of myths are deflated. Examples include that: Christianity was responsible for the demise of science; the medieval church taught the earth was flat and prevented human dissection; Galileo was imprisoned and tortured by the Inquisition; Newton's cosmology eliminated the need for God; Darwin reconverted on his death bed; and creationism is uniquely American.

While there may well be other myths about science and religion that could be exposed, *Galileo Goes to Jail* contains a comprehensive list. I found that I still held several of the myths torn down in the book.

Each case is well argued. In some cases, like the titular "Galileo Goes to Jail," there exist detailed correspondence and court records. It simply was not possible in Galileo's time in Rome for the 69 year old to have been removed from luxurious ambassadorial apartments to a dank cell in a torture chamber and then be well enough to receive sentencing in the well documented few days he was in the Inquisition's hands.

At the other extreme there is very little detail available for the Huxley Wilberforce debate. What exists from eyewitnesses either paint it as not decisive or to reflect the observer's prejudices. Nonetheless the myth that this was a major win for the evolutionary Huxley developed soon after the debate.

Another strength of the book is that it is nonpartisan. The contributors include Christians, Muslims, a Jew, atheists and agnostics. The common motivation is a commitment to understanding the truth.

I cannot commend this book highly enough.

Robert Brennan
Wontulp-Bi-Buya College
January 2022

James Ungureanu: *Science, Religion, and the Protestant Tradition: Retracing the Origins of Conflict*

Pittsburgh: University of Pittsburgh Press, 2019; 309 pages. ISBN-13: 9780822945819.

One of the most tenacious views of the relationship between science and religion is the conflict myth. I first met James after he had been granted permission to go through John William Draper's personal papers at the US Library of Congress. The chemist Draper was the author of the *History of the Conflict Between Religion and Science*. Draper along with historian Andrew Dickson White, the author *A History of the Warfare of Science with Theology*, are often accused of originating the conflict myth. In this book Ungureanu closely examines Draper and White.

Ungureanu's work is meticulous scholarship which explores both important detail and avenues of investigation often not seen in the historiography of science in previous decades. The careful study of correspondence and papers affords the opportunity to gain clearer detail about the motivations and actual arguments of key people and the circles of influence of which they were part. Ungureanu's book is important for all serious scholars of the history of science and for those wishing to gain a clearer understanding of the long relationship between science and religion.

Ungureanu begins with two chapters that detail the lives and thought of Draper and White. Surprisingly, he demonstrates that both men, rather than being dogged champions for science over and against religion, firmly believed in the importance of the Christian faith rightly interpreted. Draper was committed to the development of history as science. White opposed what he and many others saw as untenable dogmatic theology and dogmatism. White's well known anti-Catholicism was directed specifically at authoritarian dogmatism.

In these matters, Draper and White fitted well into the mood of Nineteenth century North America. They were broadly received among academic and lay people. Their influence extended to Europe and beyond. The second surprising conclusion is that Draper and White merely repeated rather than originated many of the myths in their histories. While their work has propagated general "knowledge" of the myths, the myths and doubts were already being expressed by many people.

Appropriately, Ungureanu's next two chapters trace the evolution of English Protestantism and the curiously North American development of liberal Christianity and theology. Many saw the nineteenth century as a time of a "second" reformation as questions raised by the new sciences of geology and palaeontology and the developing school of higher biblical criticism were raising serious and seemingly irresolvable questions about traditional faith. Draper and White were two of many who tried to bring about reconciliation and rational revision of theological thinking. A who's who of Nineteenth century and earlier scholars influenced them. These names included people as widely read as Lyell, Asa Grey, Darwin, Priestley, Schleiermacher and Hegel across geology, biology, chemistry and theology.

Ungureanu's last chapter demonstrates how Draper and White's commitment to reconciling science and a revised liberal protestant theology failed to satisfy liberals, conservatives and unbelievers. Indeed, many American readers believed White's conciliatory approach was a ruse. Noting that Draper and White's publishers had their own secularising agenda it becomes no surprise that Ungureanu concludes, "At the beginning of the twentieth century, rationalists, freethinkers, secularists, and atheists seized up the historical narratives of Draper, White and other liberal Protestant historians and theologians, adopting them as weapons in the campaign to extinguish all religion."

The penultimate fifth chapter explores the role of Draper and White's New York publisher Appleton. Their editor Edward Livingstone Youmans actively promoted Draper and White's books and became founding and long-term editor of *Popular Science Monthly* (this journal was sold in the 1920s and evolved into *Popular Science*). Youmans was a strong seculariser who emphasised conflict and played down conciliatory efforts between science and theology. While Draper and White's writing regularly appeared in the journal's pages, their conciliatory approach to religion was largely downplayed.

What marks this book out is Ungureanu's attention to how widely Draper and White's works were read. The answer is they were extremely widely read, generating much discussion in newspapers and other media. Because Ungureanu is able to trace the extreme breadth of the readership of Draper and White's work and the readership of the journal, it becomes apparent why the conflict myth became ubiquitous.

This book is an important contribution to our understanding of the development of the history of science and its relationship with religion. I highly recommend this book for clearly explaining the nuanced and complex relationship both Draper and White had with religion, theology, science academia and the general public. Ungureanu's attention to detail as well as exploring how widely their work was read by the general public and the considerable influence of their editor's opinions demonstrates good technique for how to write history.

Robert Brennan
Wontulp-Bi-Buya College
January 2022

Ronald E. Osborn: *Death Before the Fall: Biblical Literalism and the Problem of Animal Suffering*

Downers Grove: InterVarsity Press, 2014; 197 pages.
ISBN-13: 9780830840465.

I must confess when I saw this book, I thought: "is there any point in reading *another* book on the creation-evolution debate? Of the making of *them* there is no end." But yes, it is worth reading, for while Ronald Osborn (Associate Professor of Ethics and Philosophy at La Sierra University) does cover lot of familiar territory, his interest in animal suffering brings an interesting perspective. Indeed, his discussion of theodicy, animal suffering and issues of creation and evolution is worth the price of admission on its own.

Osborn's interest in literalistic biblical interpretation and the problem of animal suffering arises out of both his Adventist faith and his experience of the world. A recollection of his childhood experience of Mana Pools in Zimbabwe capture his primary concern well: "All around us was a world that was deeply mysterious, untamed, dangerous, beautiful and good, waiting to be explored. And the danger was part of its goodness and its beauty" (p. 13). The quest to reconcile the realities of that world, with all its death and suffering, with belief in the good Creator of Scripture lies at the heart of this book.

The bulk of the book falls into two unequal parts. The first, and longer, part, *On Literalism*, deals with literalist readings of the creation accounts and their attendant rejection of the science of evolution and deep time. Its nine chapters begin with a "plain" reading of Genesis 1–3 (by which he means a *literal* but not *literalistic* interpretation), before identifying the central problem that literalists seek to address. He then shifts gear to address more general matters relating to the philosophy (and theology) of science, and the problems they raise for biblical literalism. This leads him to a useful discussion of the sociology of knowledge and literalism and what he calls (somewhat misleadingly) their "gnostic" emphasis on special knowledge, available to the select few of the "inner circle" of true believers. Part one closes by appealing to four major figures of Western biblical interpretation and their non-literalistic interpretations of creation texts, before arguing that we need to escape the trap of modernist foundationalist epistemologies for post-foundationalist critical realism. Much of this is familiar material, if handled in an interesting and engaging manner.

Part two, *On Animal Suffering*, while shorter at only four chapters, addresses Osborn's principal concern, and deals with material that may be less familiar to folk with an interest in the interaction of science and Christian faith (and so I will say more about it). He opens by showing that while evolutionary creationism has a problem reconciling natural evil with the goodness of God, biblical literalism has a more acute—and theologically damaging—one. Whereas the former can account for it by way of permitted results of a creational "free-processes," the latter can only account for it by direct divine will, creating serious theological dilemmas. Of these, the matter of the "curse" is the most theologically telling. For this requires that all non-human death, suffering, predation, etc., is deliberately intended by God as both punishment on human sin and instruction to turn us back to God. This prompts him to ask somewhat acerbically: "What would we think of a parent who decided that the best way to educate their child in the combustibility of fire was to place the family cat on the stove? The child might learn something about fire, to be sure. But what would they learn about their parent?" (p. 138).

He moves on to a brief consideration of "C. S. Lewis's Cosmic Conflict Theodicy," which suggests that at least some animal suffering is the result of the work of (the) Satan who now, in some sense, "rules" over earth. While he is sympathetic to such an argument, he notes that we don't find the biblical authors accounting for animal suffering in these terms. Rather, we see an awe-filled realisation that predation is part of God's good world and God's governing of it. In Job in particular, God revels in the world as it is in all its terrible beauty and power: "The God of Job is not a God who delights in defanged lions" (p. 154). But even this, for Osborn, is not enough: "there remains a deep scandal in death and suffering in nature... [t]here are things under heaven and earth that we should not be at peace with" (p. 157). His response is to adopt a *kenotic* and radically Christocentric reading of creation (akin to that of Polkinghorne and others). His final substantive discussion deals with an experientially-anchored theology of sabbath in which he presents an economics of abundance and sufficiency that stands in stark contrast to late-modern capitalism and industrialised food production that brutalise both the environment and the animals we consume. This, he argues, raises questions not about God's goodness and justice, but ours, and calls us to penitent action, not mere theologising.

As can be seen, there is much to glean from *Death Before the Fall*, not least of which are some lovely personal reflections and memorable quotes. But one of his stated aims is to deal with these issues with charity, rather than adopt the theologically-more-enlightened-than-thou attitudes towards our benighted literalist sisters and brothers that so often bedevil this debate. He is at best only partially successful, and often slips into rhetorical excess. For instance: "Is it in fact the authority of Scripture in all of its richness, power, and often enigmatic and untameable diversity that we are being asked to be faithful to? Or rather a rigidified mental system and the unquestionable authority of its self-appointed guardians at any cost?" (p. 58). Perhaps it is naïve to think we can have friendly discussion on such contested matters; but language like this seems to be aimed at eliciting cheers from the choir, rather than prompting reasoned debate.

There are also several places where his work would benefit from greater conceptual clarity. For example, he presents his treatment of animal suffering as a *theodicy*. But strictly speaking, what he offers is a *defence* (a *possible* explanation of why God might permit evils of this kind), not a *theodicy* (a positive presentation of the proposed reasons why God permits or wills evils of this kind). Drawing that distinction would make his discussion both clearer and more cogent. Aspects of his kenotic theology of creation are also open to question, especially his granting priority to new creation over the original creation rather than recognising that Christ affirms and vindicates the original creation (so, Oliver O'Donovan, *Resurrection and Moral Order,* Eerdmans, 1994).

Overall, many will find that this book breaks little new ground. But Osborn's approachable treatment of the philosophical and theological problems associated with biblical literalism make this a helpful book to put in the hands of some of your enquiring friends. Moreover, it does present important reflections on animal suffering and how we should—and should not—account for it in light of our belief in God as creator. On that account, it is worth reading.

Andrew Sloane
Morling College
January 2022

Graeme R. McLean: *Ethical Basics for the Caring Professions: Knowledge and Skills for Thoughtful Practice*

London and New York: Routledge, 2021; 240 pages.
ISBN-13: 9781032009582.

Professional ethics may be conceptualised in two quite different ways. It may be understood as an exercise in the application of general ethical theory (ies) to the particular issues which arise in that profession. Alternatively, clinical medicine, nursing, social work, and so on may be understood to generate their own "internal" morality: a complex of professional role-generated norms and commitments, which arise from the nature of that activity with particular and characteristic ends. The latter understanding may be derived from virtue ethics, specifically the work of Alasdair MacIntyre on moral practices. This kind of understanding might provide doctors, for example, with particular reasons for believing that killing their patients is wrong for doctors, apart from any general considerations of the wrongness of killing.

Graeme McLean's book is based primarily on the first approach. He writes as a philosopher (he is Adjunct Research Fellow in Philosophy at Charles Sturt University, and an ISCAST Fellow), albeit one with experience in teaching healthcare students and who incorporates the insights and perspectives of healthcare professionals. His premise is that philosophy helps us to think critically and that "academic philosophers offer very helpful answers to some of those difficult questions (about how we ought to act)" (p. ix). But right up front he also recognises the objection the reader will undoubtedly raise, namely that philosophers themselves disagree about the answers to some of the most difficult questions in healthcare. Nevertheless, he is confident that philosophy offers helpful tools for critical reasoning that all can and should employ.

This is one of the book's strengths. McLean discusses the basic tools of logic by which one can determine if an argument is first, valid, and second, sound. He goes on to a technical discussion of types of argument and objections to them. Throughout, he uses examples from the philosophical literature on abortion. I wondered if it would have been better to choose a less controversial and emotive topic if he wanted the reader to engage in cool reasoning. Further, very few healthcare professionals actually perform or are involved in abortions and so a decision about its morality is not relevant to most healthcare professionals *qua* healthcare professionals.

McLean says that one way to assess a premise of an argument is through considering the consequences if it were true. If the consequences are judged "false" or "unacceptable" then "the view cannot be right" (p. 38). But false or unacceptable to whom? Assessments may vary. But McLean believes in many cases there will be agreement, based on what he has stated explicitly is his own starting point for ethical reflection—that ordinary people possess some "common moral sense." But how far this will get us in seemingly intractable debates is questionable, as he himself admits in the book's later chapters, where he says this intractability arises from different fundamental convictions or worldviews.

McLean uses the "four principles" approach of Beauchamp and Childress (principlism) to examine a series of case studies in medicine and social work. This is another very helpful section and could be used as a basis for group discussion. Given a general audience with, presumably, a wide range of worldviews, including religious commitments or lack of them, this approach is reasonable, as—although among academic bioethicists its inadequacies are widely recognised—no alternative has captured widespread acceptance.

One of these inadequacies is that it requires to be supplemented with virtues. Indeed, McLean acknowledges what many bioethicists seem to ignore, that most "hard" cases are not hard because it is difficult to know what is right to do, but because it is hard to do what is right. What is required, he says, is a commitment to the principles—what I would call virtue, the qualities of character that motivate one to do the right thing.

Another limitation acknowledged by the author is that the principles (as Beauchamp and Childress themselves acknowledge) often need to be balanced against each other, but themselves provide no guidance as to how they should be balanced. This will depend on the circumstances of individual cases.

But there is another limitation of the four principles approach that McLean does not mention. There is an inherent inconsistency between the claims that such a morality is, on the one hand, shared by all reasonable persons—a matter of "common" sense—and, on the other, based on conventions and traditions. The particular tradition in which principlism arose was U.S. individualism, and it does not always travel well beyond this context. One way to make the four principles less culture-specific is to regard them simply as a framework or initial mapping for approaching individual cases. This is how McLean uses them. But then it is apparent that their content (specification) must come from elsewhere. As must the basis for balancing the principles. One needs a thicker concept of the good than the principles provide, in order to apply them.

In Western liberal individualist societies, that reject the promotion of a "thick" concept of the good, inevitably the principle of respect for patient autonomy becomes dominant. There is no account of the good to give content to beneficence other than the patient's own conception of the good for him or her. But then it seems we have lost altogether any sense of the healthcare professional as a moral agent. Rather they become technicians or "service providers" implementing the patient's wishes. The patient or client becomes a consumer. Healthcare becomes "values free." We have moved a long way from the idea of professional ethics with distinctive values and codes of conduct.

Clearly Mclean does not advocate this view. He speaks of the distinctive features of professional ethics and the traditional definition of a profession that involves particular standards of conduct. And he speaks of the power imbalance in the relationship between a health professional and their patient/client that involves the need for trust. In the discussion of euthanasia, he refers to the duty of care which overrides even a patient's expressed wish to be assisted to die.

Ethical Basics for the Health Professions deals much more sympathetically with a religious worldview and a religious, specifically Judeo-Christian, understanding of ethics than most standard bioethics texts. There are substantial sections outlining this view and specifically its understanding of the human person in the discussion of antenatal screening for foetal abnormalities. It is contrasted with a utilitarian view exemplified by Jonathan Glover and Peter Singer. The book closes with a challenge: "What kind of carer will you be?"

I would recommend this book, especially for Christian healthcare students and practitioners.

Denise Cooper-Clarke
Ethos Centre for Christianity and Society
February 2022

Graeme R. McLean: *Ethical Basics for the Caring Professions: Knowledge and Skills for Thoughtful Practice*

London and New York: Routledge, 2021; 240 pages.
ISBN-13: 9781032009582.

It seems almost impertinent to offer some additional observations on Graeme McLean's *Ethical Basics for the Caring Professions*, following a more technical earlier review by Denise Cooper-Clarke. I do so only to add some special emphases concerning the accessibility and relevance of the book to the "ordinary" layperson like me.

While Denise does indeed recommend the book, "especially for Christian healthcare students and practitioners," I thought that perhaps her high-level analysis with reference to wider philosophical considerations might nevertheless deter a potential reader, for whom the book would be entirely appropriate. I am sure the book would be helpful to many people who, by force of circumstance, are confronted with the same (and indeed other) ethical issues and would benefit greatly from the tools for thinking about them which the book offers. The author's own words in a footnote on the last page seem to agree: "As is my strategy throughout this book, I am not trying to provide a scholarly overview of the topic, but rather to expose and discuss the *basic questions* that I believe the caring professions need to face." [author italics]

The issues McLean engages with go to the heart of what it means to be a society that values people, and cares for them accordingly.

It is not sufficient justification to do something to, or even *for* someone, medically, psychologically or otherwise, just because we *can*. McLean calls into question some of the justifications which are currently offered for practices which, even if legal, should be profoundly disturbing to a person of good conscience, let alone a person of Christian faith. The end alone doesn't justify the means.

At this point, perhaps I should declare a personal interest. I was privileged to proofread the manuscript, not primarily as a long-time friend of the author, in which description I unashamedly rejoice, but as someone who might to some extent represent the intended readers—people of average intelligence, and without finely honed logical skills, many of whom are nevertheless faced in the workplace sometimes with profound ethical issues which they can't avoid, and with which they need help. That is to say, I could hope to act as a sort of litmus test of what would be clear to such people, and what might be clearer if it were stated differently. I have no vested interest, but only the desire that people in the caring professions, as well as carers and citizens in general, should be strongly encouraged to read it.

For me, an outstanding feature of the book is that I could follow the argument! Indeed, I was engrossed by it, even though I am not a health practitioner, but simply a Christian wanting to think straight about such things. Indeed, in this case the issues have profound implications both for the health professionals *and* for those they care for. And not just professionals either, but ordinary people who may be faced with decisions on behalf of family members who may not be able to speak for themselves, as well as pastors and teachers to whom people look for wisdom in such things.

It was no surprise that I found the book thoroughly accessible, since Graeme has demonstrated a life-long gift as a brilliant teacher and communicator—an ability that manifested itself in earlier years as he held senior secondary students at summer camps engrossed in dialogical teaching about Christian beliefs, and as a tutor at Monash University. In his subsequent teaching career he received awards as the outstanding teacher at not one but two universities (the University of the Witwatersrand in South Africa in 1997 and Charles Sturt University in 2008), as well being recognised by the Australian Learning and Teaching Council in 2009 for his "outstanding contribution to student learning." I cite these things briefly not to give him an unnecessary or gratuitous compliment, but simply to underline the point once again that the "ordinary" reader can expect to get value from this important book.

Whether oral or written, Graeme's teaching is characterised by laser-like clarity, thoroughness and passion, as well as patience and empathy. This book is no exception, but rather a shining example of those characteristics. It is in two sections. The first section has five chapters which set out an invaluable framework for thinking straight—logically—about truth claims in general, and ethical issues in particular. The second section applies this way of thinking to the issues of euthanasia (Chap. 6) and disability, screening, and the value of human life (Chap. 7).

No dry academic tome, it is liberally sprinkled with examples and stories that illustrate the argument at every turn. Indeed, it is at times entertaining, as the author exposes one fallacy or another, or comments on experiences with students, with dry wit. And finally, those who enjoy *good writing* will not be disappointed. But that's the icing on the cake.

Some time ago I was talking to a friend who is now retired after a very distinguished medical career, and who has been actively studying the spread of "voluntary assisted dying" (VAD) worldwide. I was talking hopefully about the potential contribution of this book to the debate, once it was published. He was pessimistic at two points: first, that the horse has bolted as far as legislating for VAD goes in many jurisdictions; and second, that most medicos "don't *want* to think" about such issues, either because they are too busy, or because it is too hard.

One hopes, rather forlornly, that he is too pessimistic. But be that as it may, this is a book which could strengthen the confidence of Christian medicos in the respectability and relevance of theistic beliefs, and reward countless others of us who just worry about societal trends without being sure how to think straight and think Christianly about them.

Finally, though, it needs to be emphasized that this is not a book that belongs mainly on the desks of Christian people. It is a conspicuous example of argument which stands up in any arena on its own merits, even as (and partly because) it neither hides theistic assumptions, nor zealously promotes them. I am excited, though not in the least surprised, that it has found a world-class publisher, and hope that it will be read by many "ordinary Christians," and perhaps even lead others to seriously reconsider the idea that "the fear of the Lord is the beginning of wisdom."

Tom Slater
November 2022

Mike Hulme: *Why We Disagree About Climate Change: Understanding Controversy, Inaction and Opportunity*

Cambridge: Cambridge University Press, 2009; 434 pages.
ISBN-13: 9780521727327.

In Mike Hulme's view "Climate Change" is now a social phenomenon. Having worked its way into our conversations, thinking, religions, community standards, and identity, it influences the cars we buy, the stories we tell our children, and our worship at church on Sundays. It is far more than just a technical issue—to him, that is "climate change"— and dealing with it is not just a series of binary choices between simple, opposing right and wrong options.

Currently professor of human geography at Cambridge University, Hulme is a climate scientist who is also a Christian. Throughout this work he argues that issues around climate change are intractable, complex, and nuanced, and that the solution is not just a matter of pumping less carbon dioxide into the atmosphere, or persuading others to do so.

The book is for a secular audience, and Hulme makes little attempt to engage with Christian theological writers on the subject. However, he does unashamedly draw on his Christianity to make his case.

Hulme takes the debate away from doomsday scenarios, targets, and deadlines, and presents the landscape of this phenomenon, inviting us to decide where we will go within that landscape on our own journeys. But in no way does he minimise the enormity of the changes and threats we are facing, or the values and approaches, hopes and expectations we bring to the topic.

Hulme starts by listing some ways "Climate Change" has been captured in the community. He names it as a battleground between different philosophies and practices of science, a justification for the commodification of the atmosphere, an inspiration for a global network of new or reinvigorated social movements, or a threat to ethnic, national, or even global security.

In each chapter, Hulme highlights a different facet of the debate that makes the issue so difficult. These include: what climate is, the role of science in responding to threats, how we value things, and our different understandings of risk.

Hulme also notes our penchant for using crises such as climate change as vehicles for pursuing our own ideals for society, nature, and a better world. However laudable our intentions, the collective result is that they have made our responses horrendously complex and messy. Illustrating this, Hulme names mitigation strategies that range from a single universal policy target, or a single international carbon market, to tackling worldwide poverty and seeking a single global policy regime. It is not that any of these are misguided, although Hulme questions whether any of them can be practically implemented. Rather, Hulme's concern is that, together they have created a "log jam of gigantic proportions," which is not only insoluble, but is perhaps even beyond our comprehension. He argues that we need to work within this complexity rather than to strive to conquer it.

Hulme insists that because of its social nature, we need to go beyond seeing "Climate Change" as a problem, to instead see it as an idea to recognise and use within society. To do this, he suggests we employ myths to help us understand our situation—here "myths" is used to describe those stories that embody the beliefs which underly our approach to everyday or scientific reality.

He sees myths as attitudes of mind that can be used to mould the idea of "Climate Change" to serve many of our psychological, ethical and spiritual needs, and promote novel outcomes in all sorts of areas such as creative arts, intellectual property, energy production, confronting poverty, and so on.

He identifies four myths that can be used to help us understand four key psychological instincts: nostalgia, fear, pride, and justice. These myths are:

- Lamenting Eden—recalling a time when we related to nature in a simpler, less ambiguous way and realising we cannot go back.
- Presaging the Apocalypse—acknowledging the fear that we are tumbling towards a future catastrophe in which the threats are not the four Riders, but humanity itself, with our voracious and uncontrollable appetites.
- Constructing Babel—driven by a desire for mastery to control (and, in this case, repair), we strive for dominion over nature, but our construction is brittle and fragile.
- Celebrating Jubilee—confronted with the injustices and disparities of power that Climate Change is amplifying, we need to call "Time Out" and do some resetting.

Personally, I would have added a fifth myth: the Suffering Servant. Drawn from Isaiah 40–55, people have lost hope through their own behaviour and are suffering. God intervenes after a time, assuring them they have suffered enough, and is offering to lead them lovingly to a new beginning.

Hulme rejects the "problem-solution" mindset of a traditional approach, so employing these myths is not a solution in the usual understanding of the word. Instead, they are frameworks or understandings to help us live with and within the new reality. He does not necessarily endorse all that might be assumed from any single one of these myths. Thus, while he categorically rejects doomsday thinking, he recognises that the Apocalyptic myth is a reality in our conversations about "Climate Change."

Hulme's work provides enough information to set my mind running and reflect on this as a different approach to climate change, as a social construct. From this, I can reflect more broadly on who we are (our fears, hopes, and expectations) and how "Climate Change" affects our thinking.

I spent my professional life struggling with environmental problems and having had to learn to live with them. Think blackberries, feral pest animals, the hole in the ozone layer, the long-term impact of the 1939 bushfires, eucalypt dieback, and dry-land salting. So, I had always thought it optimistic that we thought that we could "solve" climate change *per se*. Throw in the international dimension and my head really starts to spin.

Hulme does not ask us to stop striving to behave responsibly on this issue, but he does reject our hopes that these will "solve" the problem and take us back to the day when this was not an issue.

In among the confusion and muddling—the frustrating international conferences, the scientists and engineers struggling to suck carbon dioxide out of the atmosphere, or any other attempts to address the problem of climate change—we should step back and seek to understand the nature of this social phenomenon. Hulme encourages us to do this.

And so, the book leaves me with thoughts and ideas tumbling through my mind about how we must now live. I would encourage anyone engaged with "Climate Change" to read this book, not to tell them how to think, not for answers, but for questions and the possibility to start thinking afresh and come up with new, innovative and helpful perspectives.

One challenge for me was to reflect on how my faith in Jesus Christ might make a difference (and, if so, how). That said, and being familiar with the myths he draws on, I can also see beyond them and on to the God who created this untameable nature, and loves and cares for it as much as God loves and cares for me. That is a myth I can live with and draw on.

Richard Gijsbers
February 2022

David Frost: *Blind Evolution? The Nature of Humanity and the Origin of Life*

Cambridge: James Clark & Co, 2020; 176 pages.
ISBN-13: 9780227177112.

David Frost's *Blind Evolution?* addresses a longstanding and still hotly debated controversy about Christian attitudes to science. His primary purpose is to carry the day in what he calls a "'boots-and-all' attack on atheistic Neo-Darwinism," especially its claim that life in all its variation has occurred solely by directionless (and in essence) meaningless chance. Grounded in a deeply held conviction that the Christian God is the creator of all that exists, Frost's aim is not only to undercut the supposed validity of these claims but even more the credibility of their chief advocate, former Oxford Professor Richard Dawkins, well-known for his scathing attacks on theistic religions, whose book *The God Delusion* sold over three million copies.

Frost's second motive is pastoral. He wants to offer his readers a more evidence-based answer to life's deepest questions, especially the question of undeserved suffering, in place of Dawkins' unbelief-fostering and hope-destroying agenda. Hence, he emphasises that science alone is not the whole truth.

Professor David Frost taught English literature and religious studies at several universities, including St John's College, Cambridge, UK, and the University of Newcastle in Australia. On his retirement, he returned to Cambridge as honorary Principal of the Institute for Orthodox Christian Studies. His research and publications centred on literary subjects like Shakespeare, seventeenth-century English drama, liturgical works like a new translation of the Psalms, and the liturgy of St Chrysostom. The book under review is Frost's first and only work in its genre. Frost aligns himself with the Intelligent Design Movement (IDM) in the creation-evolution debate.

A quick look at his text reveals some intriguing clues about his approach. Frost does not analyse data or explicate published scientific papers but offers mostly anecdotal evidence in keeping with his view that truth can also be found in poetry, dreams, visions, and, of course, in the Christian scriptures. Lengthy expository passages alternate with reports of dreams and personal emotional states amplified by liberal use of poetry, biblical quotes, descriptions of photographs, occasionally laced with satire and ridicule towards his interlocutors. The book features nearly sixty photographic illustrations and one full-page cartoon, which shows Richard Dawkins on the lookout of the "Ship of Fools" on a perilous voyage, with sails in tatters, flying a pennant saying, "There is no God." Other caricatures onboard depict the Pope (asleep), Anglican, Catholic, Orthodox, another clergy, a philosopher, a scientist, and a meditating Buddhist. Similar sarcasm is reserved for Charles Darwin: "How can a competent biologist like Darwin be so daft?" (p. 31). His composition engages a blend of styles, ranging from journalistic to lyrical and novelistic, all in the service of the book's announced thrust of defeating his opponents, often leaving it to the reader to discover the connection.

There is much in this book a thoughtful Christian can agree with. The radical materialism of Richard Dawkins deserves to be debunked by calling it out in its many disguises, including the hard-line Darwinism Frost seeks to falsify. To that end, he engages first in the standard rebuttals that materialist evolution is insufficient and misleading to explain, let alone generate features necessary for life like novelty, complexity, ecology, regeneration, etc. His second line of attack consists of his (or IDM's) claim that "what looks like design is designed" inferring that a supernatural agency or "intelligent designer" provided the blueprint and assembled the elements we observe as "irreducible complexity" in nature. Proponents of the IDM claim that specific complex biological structures and functions, like the eye, are best explained by an intelligent designer, not by the seemingly directionless process of natural selection.

Frost rightly argues that, notwithstanding its many benefits and vast explanatory reach, the scientific method (which he endorses as a God-given tool to discover how the world works) is nonetheless limited to what it can describe and measure. Therefore, science alone cannot fully account for reality, contradicting the materialist claim that science is the whole truth.

According to Frost, what is lacking is a more thoroughgoing understanding of "truth" that should include another, much older form of knowing, the knowing by intuition, insight and inspiration, by anecdotes, poetry, prayer, prophecy, and (biblical) revelation —*nous* in Greek. Against the claim of a materialist culture that rejects this kind of knowing as unverifiable, Frost advocates that the scientific method plus *nous* is the only reliable path to the "whole truth" (about God and the creation) and thus for trust in the goodness of God when faced with inexplicable circumstances and undeserved suffering. From a Christian perspective, these are highly commendable aspirations, albeit not without their own hermeneutical problems which Frost leaves untouched.

This outline of Frost's approach to delivering what he considers his knock-down case against Dawkins and Darwin prompts the question of whether or not he has achieved his aim, which in his own words includes the collapse of Darwin's "house of cards."

Frost has chosen the right target to attack a worldview that offers only hopelessness (because its universe and its human inhabitants are nothing but chance-driven, mindless matter). However, going to battle by setting one belief structure against another can only lead to a clash of two sets of mutually exclusive presuppositions and thus to a futile escalation of hostilities. Throughout the book, Frost ties Darwin's work to the twentieth-century version known as Neo-Darwinism and Dawkins' atheistic dogmatism. This unfortunate conflation has three troubling effects: it renders "evolution" synonymous with the promotion of atheism; it takes Frost close to committing a severe category error; and it gratuitously revives the outdated "combat model" of the theology/science discourse.

Frost's a priori rejection of evolution makes him miss the point that Darwin's concepts, although not fully developed at their original formulation, represented a paradigm shift in developmental biology that continues to bear fruit. The scope of "evolution" is still evolving, representing today a global, multi-disciplinary research programme of ever-widening purview ranging from genetics to consciousness studies. Frost also omits to mention that the biological sciences themselves have begun to break with traditional Darwinism by distinguishing between mutations and natural selection while acknowledging that environmental pressure has been found to work on both in its own way.

Lastly, Frost offers Intelligent Design as the only plausible explanation for the complexity and variety of the natural world. Although the implied "designer" is not necessarily the Christian God in this conception, Frost forges a close link nonetheless, inviting the theological objection that the doctrine of God as Creator calls for a more far-reaching description. Karl Barth, for instance, devoted two thousand pages in his *Church Dogmatics* to the doctrine of creation.

More broadly, Frost sets out to address questions of ultimate reality, yet his view of evolution remains narrowly earth-bound. What seems to escape him is the cosmic dimension of the evolutionary paradigm that has guided scientific inquiry already for several decades across a widening spectrum of disciplines. Frost's narrow conception affirms a deep-rooted Christian view that the creator brought forth finished products (the earth, animals, humans, stellar objects), blinding us to the *process* nature of creation. Yet, modern science discovers it at multiple levels of analysis, from quantum states to biochemistry and beyond. Here we find contingent possibilities instead of individual entities as IDM and some Christian traditions claim. There is even a biblical warrant for such reading if the multiplicity of Hebrew meanings in Genesis 1:1 is given its due. If this conception were adopted, mutations can no longer be called blind, meaningless, and directionless. Instead, they can be understood as contingent and exploratory structures of adaptation that exist (theologically) in response to the divine promise of an ultimate consummation.

In conclusion, Frost's commendable campaign against the militant materialist interpretation of developmental biology for all its strengths turns out to be less watertight than his purpose statement first suggested. Remaining wedded to the combat model, Frost also deprives the Christian imagination of the possibility of rising above the cramped frame of Neo-Darwinism, holding Christianity back from the urgently needed creative interaction between the data of modern science and the insights of Christian theology.

Peter Stork
March 2022

Ian Hutchinson: *Can A Scientist Believe in Miracles? An MIT Professor Answers Questions on God and Science*

Downers Grove: InterVarsity Press, 2018; 288 pages
ISBN-13: 9780830845477.

I have to admit to being slightly mystified by the title of this book, expecting the author to spend his time analysing the place of miracles in the Bible in general and Jesus' ministry in particular. This is partly true, but the subtitle is far more helpful: *An MIT Professor Answers Questions on God and Science*. Ian Hutchinson is Professor of Nuclear Science and Engineering at MIT, and for many years has been involved in the Veritas Forums in the United States. The aim of these is to provide an opportunity for university students of all beliefs, from Christian to atheist, to question established thinkers on a wide array of challenging issues. The expectation is that the speakers will provide academically rigorous input to whatever questions are thrown at them by the audience. The big questions of life are addressed in the context of the academy, something not usually found in the regular academic curriculum.

This is precisely what is provided in this book by Ian Hutchinson, who is English and a Cambridge University graduate, but who has spent the bulk of his academic career in the United States. The book contains his answers to over 200 questions asked of him in the Veritas sessions. Their subject matter is nothing if not broad. Won't science eventually explain everything? Isn't faith opposed to critical thinking? What reasons are there to believe in God? What is the difference between science and scientism? Is there intelligent design? What are miracles? Does the Bible teach science? Is God's existence a scientific question? Could a good God permit so much evil?

The result is a superb expression of a deeply committed scientist bringing to the table the wealth of his scientific expertise, all with openness and unabashed honesty. He is prepared to tackle everything raised by the students, and he refuses to duck even the most difficult and opaque of issues. He is prepared to say when he does not have what he thinks is a satisfactory answer, and also when he thinks there is no ready explanation. For instance, the problem of evil inevitably raises its head, and while he is prepared to be agnostic on some aspects of this, he does not allow atheists any slack and points out that evil creates perplexities for them as well.

The book is worth reading for Hutchinson's understanding of the way in which science works; he has a deep appreciation of the power of the scientific method and also of its limitations. This sets him up very well for seeing how science and faith interact, and the boundaries of each. This is also exceedingly useful when he delves into evolutionary thinking, archaeology, the age of the earth, and the long history of the universe. He is at ease in knowing where the boundaries of science are to be drawn, and in sketching the demarcations between scientific evidence and the mechanisms of science, and philosophical speculation in the guise of scientism. Time and again what comes through is his clarity of thinking and his common sense. His chapter on cosmology, the multiverse, and intelligent design is a model of his approach, unencumbered as it is by metaphysical and speculative theorising. So is the chapter on miracles.

The author is equally critical of the use made of substance dualism by some Christian writers, as though the Bible teaches the survival of immaterial souls. For him the Bible teaches the resurrection of the body, and that our life with God will involve a new embodiment as opposed to the persistence of a separate soul. His solid grasp of philosophy helps him resist the inroads of determinism and reductionism, and allows him to steer clear of any suggestion that human beings are "nothing but" physical machines. His critical and analytical analysis of neuroscientific issues allows him to clear away a great deal of what is so often confused and obfuscating in the views of a host of other writers. His everyday faith comes through at all points, preventing him from becoming arid and unduly theoretical.

The format of the book allows Hutchinson to provide many gems, such as the hindrance created by individualism for both science and religion. The book is worth reading for some of these gems, as it is for Chapter 6 on scientism, and the importance of non-scientific knowledge, with its final sentence: "science in [sic] not all of real knowledge; nor is scientific evidence all of real evidence."

He is never afraid to say that he might be wrong, and that this is a call to repentance and acknowledgement of one's sins. How delightful to hear this in the midst of such an erudite series of expositions, especially with his explicit acceptance that both scientists and Christians can be wrong and even charlatans. Throughout though, he strongly advocates for the place of religion in a world dominated by science, particularly when religion should be contributing to moral debate.

One area not covered by the questions and responses in the book is biomedicine and bioethics. This is perhaps inevitable for someone whose expertise is in physics and the general science-faith domain. It is also good that he does not get entangled in the fraught controversies around the embryo. He does mention embryonic stem cells but says little about them.

The vast span of topics covered in this book might be seen as a disadvantage, since any one reader will not wish to go into all of them in depth. Nevertheless, this is a minor quibble for what is an outstanding example of the best scientific thinking by someone with a thorough grounding in the Christian faith and theological thinking. Oh, for more scientists like Hutchinson who bring the best in scientific attitudes and approaches to their Christian faith, and are not beholden to the straightjackets imposed by some conservative Christian mindsets.

The ideal readership for this book is advanced undergraduates and postgraduates, plus anyone else wanting serious analyses of questions facing the contemporary faith community. The book could well be used as a pre-evangelistic tool because Hutchinson clears away a great deal of garbage that clutters thinking in today's universities and churches.

D. Gareth Jones
University of Otago
April 2022

Simon Conway Morris: *From Extraterrestrials to Animal Minds: Six Myths of Evolution*

West Conshohocken, PA: Templeton Press, 2022; 272 pages. ISBN-13: 9781599475288.

Simon Conway Morris is a Cambridge palaeontologist best known professionally for his work on the Burgess Shale and the Cambrian explosion. His less recent works include the books, *The Crucible of Creation* (1998) about the Burgess Shale fauna, and *Life's Solution: Inevitable Humans in a Lonely Universe* (2003) about the role of convergence in organic evolution. Conway Morris is a Christian and was keynote speaker at the ISCAST Conference on Science and Christianity (COSAC) in 2009. More recent works have included several published by Templeton Press, including *The Runes of Evolution: How the Universe became Self-Aware* (2015).

Conway Morris's latest work is subtitled *Six Myths of Evolution*. These are the common myths in the public mind, the media, and even among some scientists: the myths of no evolutionary limits, the essential randomness and directionless-ness of evolution, the essential role of mass extinctions in clearing the ecological decks, prevalence of missing links, animal minds as precursors of human minds, and the supposed abundance of extra-terrestrial intelligence. I will summarise each of these sections in turn.

The "no limits" myth is very popular in the public mind. "Life will find a way" is the mantra whenever the potential limits of life are raised in discussions. But will it? In this chapter Conway Morris shows that there appear to be "Great Walls" in biology; these walls are set by physics, chemistry, and information requirements. Furthermore, while in theory almost everything may be possible, only a few patterns seem to work in the real world, driving to multiple independent re-appearances of the same solution. Although majoring on this particular myth, rather confusingly this chapter begins with refuting a myth Conway Morris does not list, the "simple to complex" myth, where he argues that there are relatively short-lived episodes of great biological innovation leading to great complexity, followed by streamlining and simplification. Conway Morris describes many examples of this process, including the rise of amphibians, birds, and mammals. This might have been worth a chapter of its own.

The randomness myth was perhaps expressed best in the essays of the late great Stephen Gould. Evolution was a random process, lacking constraints, beyond the bounds of minimum complexity. There was no direction to evolutionary processes. However, as Conway Morris has already documented extensively in earlier works such as *Life's Solution* (2005), similar features appear independently again and again throughout evolutionary history to solve the same problems in different taxa. This phenomenon occurs at all scales, from the convergence in body shape in sharks, dolphins, tuna, ichthyosaurs, and mosasaurs, down to the fundamental molecules of life such as oxygen-transporting compounds, optical sensors, and chlorophyll. Again, while in theory almost everything may be possible, the reality that only a few patterns actually seem to work provides a major constraint on evolutionary patterns and a strong driver for specific solutions to biological requirements.

Mass extinctions are those great events in the history of life where large numbers of taxa, often at high levels, disappear over a short period of time. The two largest, and the two that have most grabbed the popular imagination, are the Permo-Triassic extinction, where perhaps 90% of species died out, and the Cretaceous-Tertiary extinction, which saw 75% of species become extinct (percentages vary). The cause of the Permo-Triassic extinction is not yet understood but the Cretaceous-Tertiary event was largely because of an asteroid impact in what is now Yucatan. Conventional wisdom argues that mass extinctions clear the decks, wiping out old groups and allowing new or suppressed groups to flourish. Had the asteroid not struck Yucatan, the argument goes, mammals might have remade little scurrying things and the world would still be dominated by dinosaurs. However, Conway Morris argues that the picture is much more complex. Many taxa that disappeared during the extinction events were in fact already being replaced, such as extensive replacement of conifers by flowering plants in the Cretaceous. Mammals would probably have replaced dinosaurs in many niches anyway, because of more sophisticated metabolisms, much as birds were replacing pterosaurs in the Cretaceous in many niches. So rather than entirely resetting the evolutionary landscape, mass extinctions had more the effect of speeding up trends already in place.

The discovery of so-called "missing links" are beloved in the popular media, the discovery of such a possible link supposedly fills forever a gap in the evolutionary record (for example the transition from fish to amphibians or dinosaurs to birds) or proves once and for all that Darwin was right. Conversely, they are loved by anti-evolutionists: their supposed absence proves that there are gaps that can never be filled; *ergo* Darwin was wrong. Conway Morris's chapter on this myth shows that, as always, the reality is very different, on at least two counts. The first is that rather than a simple transition from sarcopterygian fish to a *Ichthyostega*-like amphibian which can be neatly plugged by a single fossil like *Tiktaalik*, what we see is a tangled thicket of forms, often with mixed primitive and advanced characteristics, where the exact line of descent is difficult if not impossible to demonstrate, even though the general relationship between sarcopterygians and amphibians is clear from palaeontology and genetics.

When I was an undergraduate biology student, animals were little more than automata, whose behaviour was the result of simple conditioning. Subsequently there have been many claims in popular science media of advanced cognition, cultural development, proto-linguistic development, and other indications of sophisticated animal minds. Conway Morris provides much-needed caution on these claims, providing counter examples and caveats that are not likely to make it into the popular media. While not the automata of my student days, animal minds remain both quantitatively and qualitatively different from human minds, according to Conway Morris. However, this chapter is weaker than the preceding ones. If humans evolved from earlier primates, as Conway Morris argues, how then did human minds appear? Are they an emergent property once a particular threshold was crossed, or was something else involved? More explanation would have been helpful.

Even weaker is the chapter on the extra-terrestrial myth. Conway Morris is correct in pointing out that while the idea of a universe teeming with sapient beings, if not entire technological civilisations, is popular, where then is everybody? Not only is there the apparent lack of evidence of technology in the observed universe, but there is also the absence of any traces of alien visitation in the billions of years of geologic time. The "great silence" Conway Morris argues, as do some others, suggests that we are alone in the universe. Life may (or may not) be common elsewhere; mind probably not. However, the reality is we simply do not know, and I suspect that Conway Morris here writes with greater confidence than is perhaps justified. The last part of this chapter becomes very odd indeed, with Conway Morris wandering into literature on paranormal and similar events. It is not clear where, if anywhere, he intends to go with this, or its relevance to the rest of the subject matter.

To conclude, this is a readable, provocative, stimulating and, for the most part, well-founded book showing how these evolutionary myths are often misleading, if not actually unhelpful. I would have liked the book to have addressed two other myths, the first that of the simple to complex. Conway Morris touched on this in Chapter 1, but a separate chapter would have been most helpful. As an astrobiologist, a chapter on the beginnings of life and the various narratives attached to that, would have been interesting. As mentioned, the last two chapters are rather weak, and the book ends on quite a strange note. Despite these caveats, I recommend it strongly to anyone interested in palaeontology, organic evolution, and to a lesser degree, the nature of mind and human evolution.

Overall, this is a useful book exploring some of the larger questions in the patterns of biological evolution by a leading thinker in the field. It is well referenced and provides much to ponder.

Jonathan Clarke
Amity University
April 2022

Graham Buxton and Norman Habel (Eds): The Nature of Things: Rediscovering the Spiritual in God's Creation

Eugene, Oregon: Pickwick Publications, 2016; 268 pages.
ISBN-13: 9781498235143

This collection, edited by Graham Buxton and Norman Habel, contains a series of essays emerging from a 2015 conference held at the Serafino winery south of Adelaide. The aim of the conference (and the essays) was to avoid consensus but to "survey the landscape with a view to intentional responsible action in caring for God's creation" (p. xx). The "Serafino Declaration" that emerged is (to me) a strange conglomeration of well-meaning but rather foggy statements about the spirituality of country, the rape of nature, the "rights" of nature, and healing of the earth as a primary mission. Very little of this is useful to those whose vocation is to care for the land and "share with justice the resources of the earth." However, for those who persist past the introduction and the declaration, there is some worthwhile content to ponder.

The bulk of the collection contains 16 essays by 18 authors. Of the authors, 15 are theologians and only three have a science background (one each from the cosmological, atmospheric, and geological sciences). Despite the conference commencing with the now *de rigueur* welcome to country and smoking ceremonies, none of the authors appear to come from an indigenous background (Australian or otherwise) or reveal any firsthand experience of indigenous culture with respect to actual living in the world. For a collection that frequently uses scientific terminology and repeatedly refers to indigenous spiritualities, the limited scientific input and lack of indigenous (specifically Christian indigenous) contributions, is something of a disadvantage.

The tone for much of the content appears in the second paragraph of the foreword by David Rhoads: "We are in deep trouble ..." This statement embraces two of Mike Hulme's framing myths for discussing climate change (and many other issues regarding the world in which we live)—the myth of a lost Eden and the myth of an impending catastrophe. The foreword also sets the tone with respect given to indigenous spiritualities and their attitude to the world. Unfortunately, this respect is not tempered by a recognition that such spiritualities are often antithetical to the Christian faith, have proved singularly unable to rise to the challenge of dealing with contemporary issues, and that many indigenous people have willingly adopted Jesus (the 2006 census reported 73% of the Indigenous population of Australia identified as Christian) rather than following their traditions.

Nevertheless, there is much of value in the essays for the persistent reader. Denis Edwards' discussion on Athanasius and the doctrine of creation is helpful in providing an insight in patristic thought on these issues. And I share a common delight with him of the South Australian landscape, especially the Flinders ranges and the Willunga region. Likewise I found Santmire's discussion of Augustine insightful in his recognition of the aspects of power and presence in the natural forces (in his case Niagara Falls); this resonated with my own experience (including a Force 10 storm in the Southern Ocean or exploring White Island crater). I passed over Gardner's linking of Elijah's "still small voice" with a "hum" heard on a remote farm and to the "Om" of Pravana yoga with a "hmm" of my own!

The chapter by Buxton is disappointing for several reasons, one of which is its uncritical acceptance of the Lynn White thesis (that Christianity is responsible for the ecological crisis) without any discussion of or even reference to the many rebuttals of that thesis. Another is its limited engagement with more-or-less orthodox Christian understanding of creation (Jürgen Moltmann and St Francis excepted), instead preferring heterodox writers (Bruno), New Age philosophers (Capra), proponents of imaginative naturalism (Eiseley and Thoreau), and (of course), non-Christian Australian Indigenous spirituality.

Mike Pope's essay nicely encapsulates why Christians need to understand a whole-of-system approach when thinking about the world (e.g., the six interacting spheres of the biosphere, geosphere, atmosphere, hydrosphere, and celestial sphere). Likewise, his is one of the few essays to talk about environmental management, albeit for only 15 lines and focussing entirely on wilderness rather than the inhabited spaces where people live and work. Pope is to be commended for referring to the work of some Christian indigenous writers, although I would like to have seen engagement with a wider range.

White's essay robustly accepts the reality of cataclysmic processes operating through the Earth in time and space and that when these interact with the human sphere the consequences are often disastrous. Unlike many of the other contributors to this book, White understands the importance of human management to minimise such consequences and has a strong commitment to Christian hope in the new creation.

Mark Worthing also focusses on the question of suffering. He touches briefly on the scope of suffering in the physical world, the suffering of God, and the atonement as God's response and solution. Perhaps more could have been done here on the eschatological dimension of the New Creation, perhaps following on from the work of David Wilkinson.

I found the discussion of the Wisdom literature as a means of understanding the creation by Habel interesting and helpful. This part of Scripture has been under-utilised in my experience, with its observations of animal and plant behaviour, space and time, place, and context. A minor frustration was his repeated use of the term "scientist" for the Wisdom writers of ancient west Asia and the Mediterranean. Insightful they were, but they were not doing science as we understand it. I also suggest that saying that the Wisdom literature portrays the Spirit as some kind of life-force, as opposed to sustaining creation, may be taking things a bit far. The following chapter by Deane-Drummond follows on the Wisdom theme and provides some fascinating snapshots of interactions both within animal social communities and between them and human communities. However, in seeing evidence for play and a sense of justice among animals she relies perhaps on too few sources as well as passing over contrary evidence. Nonetheless, with her I agree that we should not hesitate to see the mind and wisdom of God at work in the biological world.

Balabanski begins her piece with a blast from the past—the "Louie the Fly Mortein" advertisement familiar to older Australians. She uses this to examine the relationship between purity and impurity in the Bible from an ecological perspective, in particular that of the microbial biome. With the COVID-19 pandemic, such reflections have taken on a new relevance. Somewhat similarly, Colgan's essay explores the cosmic imagery of Jeremiah 31:35–37. In it she sees both the vast and incomprehensibly complex nature of the creation and at the same time its interconnectedness. Colgan may be going too far in suggesting this provides an emerging ecological perspective of creation, but she has nonetheless brought an oft-overlooked yet beautiful and powerful passage into our understanding of God's world that will repay much future reflection.

Liederbach links worship with stewardship. This is important because it returns the focus to God, avoiding the excessive focus on the created order of some eco-theologies both inside and outside the evangelical community. True worship includes caring for God's world. He finds a helpful example of this in Tolkien's description of the exploration and development of the Glittering Caves. Along the way Liederbach eviscerates (his word) the Lynn White thesis, that the root of environmental problems lies with Christianity.

I found several essays in this book less helpful. Fox's piece on the relevance of the spiritual exercises of Ignatius of Loyola was opaque to my thinking, while Rayson and Lovet's contribution perhaps draws too much on limited links between the thought of Bonhoeffer and Gandhi and, at least in this chapter, skims over the immense differences between Christianity and Hinduism. Likewise, the final piece by Elvey was valid more as a personal reflection than providing a wider understanding. But in all these the fault may lie with the reviewer.

In conclusion, I found this collection mixed. Some contributions are very good, others somewhat good, others either not helpful or even opaque. This is perhaps inevitable given the eclectic nature of the writers. The book misses out in some places by an uncritical endorsement of indigenous perspectives of the world, ignoring the fact that these have also led to massive environmental impacts over time and are in many cases simply not relevant to living in the 21st century. It also ignores how indigenous perspectives have been transformed by the gospel; it seems egregious to me when Christian theologians, harking back to pre-Christian ideas, ignore this transformation. Especially lacking for the most part is the absence of the practitioner's perspective. Conferences such as the one which gave rise to this collection can only occur through a vast infrastructure of technologies such as communications, transport, power, water, agriculture, and the rest. To speak of justice for the environment and the poor, without considering how this is to be achieved, is to tell only half the story.

Jonathan Clarke
Amity University
April 2022

Robert John Russell & Joshua M. Moritz (Eds): *God's Providence and Randomness in Nature: Scientific and Theological Perspectives*

West Conshohocken, PA: Templeton Press, 2018; 388 pages.
ISBN-13: 9781599475677.

This volume contains ten essays that were developed out of the "Scientific and Theological Understandings of Randomness in Nature" project, funded with a grant from Calvin College. The contributors come from a scientific and/or a theological background, with some papers of a more scientific orientation, others of more theological orientation seeking to respond to the scientific evidence.

The first three chapters, Section 1, focus on scientific questions: the interplay of randomness and necessity in scientific understanding (George F. R. Ellis); the universal nature of the laws of physics (Robert E. Ulanowicz); and theories of a multiverse (Gerald B. Cleaver). The chapter by Ellis which opens the collection is particularly technical in its initial discussion of quantum mechanics before launching into a more wide-ranging discussion of chance, necessity and purpose in chemical and biological systems with an eye for the question of emergence of complexity. Ellis provides a thorough coverage of the issues involved, and is the most substantial of these essays. The contributions by Cleaver and Ulanowicz are more accessible for a non-expert. Ulanowicz challenges the absolute claims of scientific laws, recognising their essentially abstract nature, while Cleaver attempts to put a positive spin on the idea of the multiverse from a religious perspective, while acknowledging the highly speculative nature of the proposal. I would add that personally I find the notion of a multiverse just too speculative and more than likely unverifiable and hence not a scientific theory at all.

The remaining seven essays, Section 2, raise the God question more fully, covering a variety of theological questions impacted upon by scientific theories: can there be genuine divine providence in a universe with randomness (James Bradley); does quantum mechanics offer a way of understanding divine action in the world (Robert John Russell); does science allow for "top down" causality and emergence, a more philosophical than theological discussion (Alicia Juarrero); the tension between classical science and belief in miracles (Erkki Vesa Rope Kojonen); the possibility of free will in a universe of necessity and chance (Veli-Matti Kärkkäinen); the challenges to freedom arising from neuroscience (Ted Peters); and how evolution requires a reframing of the theodicy debate (Joshua M. Moritz). Many of these authors are "bi-lingual," at home with both theological/philosophical and scientific literature, some more technical in nature (Juarrero), others more theological (Kärkkäinen).

There is much to learn from all these essays, though it helps to be a bit bi-lingual to get the most out of them. The first three scientific essays in Section 1 are theologically light, but overall the essays in the second half of the work display a solid, if not technical grasp of the scientific issues involved. It takes a type of courage on the part of those who are theologically trained to venture into the fields of science and vice versa, but the rewards are worthwhile.

That said I would like to make some theological criticisms of the second half of the work. The authors are generally from a Protestant background and so there is little engagement with the Catholic tradition and its insights. The questions around chance, necessity, free will and providence are as old as Christian theology, and a relatively stable solution to the questions posed found some semblance of a response in the work of Thomas Aquinas, whose name does not appear in the index but does receive some discussion in the essay by Russell. Aquinas provided a metaphysical analysis of these questions so the question is then whether such a metaphysical approach is independent of different scientific outcomes. In other words what is the relationship between metaphysics and physics? This question becomes most acute in discussions in relation to God's "intervention" in the world. So often these questions appear to place God as one agent operating among others rather than as the primary agent of all being—the scholastic distinction between primary and secondary causation. I would suggest that a more coherent understanding of the implications of what it means for God to be primary cause would resolve many of the issues raised in the second part of the work.

Nonetheless, this volume is a worthwhile contribution to the debate and there is much to learn from the various authors. There is presumed knowledge in a number of the contributions so this is not a book for a novice in the area. But for those with some degree of literacy in both science and theology it is worth a read.

Neil Ormerod
Sydney College of Divinity
July 2022

Suzie Sheehy: *The Matter of Everything: Twelve Experiments that Changed our World*

London: Bloomsbury, 2022; 317 pages.
ISBN-13: 9781526618955.

I am a theologian—that is who I *am*. It is a calling at whose heart lies curiosity—about life, the universe, and everything.

Mildura-born Dr Suzie Sheehy is an Oxford- and Melbourne-based particle physicist and science communicator—that is who she *is*, given the evidence of this wonder-full book. And at the heart of her calling lies curiosity—about what makes up the universe, "the matter of everything."

Modern physics and classical theology may seem to have little in common. Yet I found reading this book not unlike working through a volume of Karl Barth or Elizabeth Johnston: beautifully organised, an unfolding development of ideas, close attention to detail, and the priority of the human over the technical—with no diminishment of the latter. Sheehy is pleasingly curious about her own discipline's story.

As the subtitle indicates, Sheehy describes twelve key experiments in physics, from 1896 to 2021 (COVID gets attention!). The chapters are grouped in three parts. The first, "The dismantling of classical physics" outlines the discovery of X-rays, the structure of the atom, and the nature of light. It should be read by anyone who appeals to common sense. The second part, "Matter beyond atoms," I found a bit tougher although I took physics at Sydney University (in the sixties). It charts the discovery of particles within the atom, uncovered by the cloud chamber, cyclotron, and synchrotron.

The third and longest part is "The Standard Model and beyond." Five chapters trace the story from "strange" particles to neutrinos to the Standard Model that since 2000 has displaced the "classical physics" of Parts 1 and 2. What does it mean that matter is ultimately waves? That dozens of particles exist, some lacking mass? That there is so much "emptiness" in atoms and space? That gigantic machines of extreme delicacy are needed to detect the most minute, most fleeting particles? Such questions and more take one's breath away!

In a 1963 physics class, I remember Professor Harry Messel coming in one day to enthuse (in his gravelly US accent) about a new theoretical idea: "quarks, boys and girls, quarks—remember that word!" Many of the experiments described in Part 3 involve quarks (proved to exist five years after that class); I now have some inkling of why finding the "Higgs boson" was so important, if little idea of what is involved! Time and again I was glad of the index to refresh my memory, and to read endnotes that spell out details of the people involved.

Sheehy gave a well-received TED lecture in 2018 in which the experiments described in this book were outlined. Many of the concepts would benefit from diagrams, while photographs of the equipment involved in the experiments would add significantly to the interest. The book would make the basis for a great television series, especially with the visual dimension this entails.

The thirteenth chapter is "Future Experiments." This draws sharp focus on two themes found throughout the twelve experiments: the importance of curiosity as well as problem-oriented research, and the centrality of people. These are where I found this book inviting explicit engagement with theology.

Curiosity, experimentation for its own sake, does not easily get budget support in science. Yet Sheehy shows how each discovery has led to unimagined enormous benefits. Carbon dating has revolutionised archaeology, for example—Australia's Aborigines, the Dead Sea Scrolls, and the Turin Shroud get a mention. In medicine, CT, MRI, and PET machines—the fruits of particle acceleration research—are now found in most hospitals. The special light that synchrotrons emanate has seen breakthroughs in botany and geology, by enabling crystalline structures to be analysed. The World Wide Web arose from the need for computer storage space beyond what was possible physically, as data multiplied. And now the search is on for "dark matter," which may form 95% of the universe. Can we even imagine any practical outcomes from this?

For theology, to be "curious"—a favourite term of the Dutch theologian G. C. Berkouwer—does not mean neglecting or disrespecting the tradition but delighting in the privilege of being able to see how it unfolds in new situations. Yet, why is so much of what passes as theology uncritical, defensive, and energised by individual morality rather than reality as a whole? Being curious about the "story" of theology can spark significant contributions to understanding what faith means for living in this globalised world.

The final chapter opens with Sheehy telling something of her story and how it shapes her work. This exemplifies a theme throughout the book: the people involved in each experiment (including many Nobel Prize winners). Sheehy brings them to life. Over 220 scientists are named, including around 30 women whose work has not always been appreciated. In a telling anecdote, Sheehy notes that the term "computer" was coined to describe the women in the 1940s who solved differential equations; when machines took over this task, collecting data became "women's work" (pp. 293–4).

Names that have stuck with me include brothers Ernest (cyclotron inventor) and John (medico) Lawrence: their working together in the 1930s led to radiation treatment for cancer. Robert Rathgun Wilson (a jack-of-all-trades "hero" of Sheehy) blends aesthetics with physics in laboratory design, raises big funds for "big science," brings people together—and does good research! Physics involves people, Sheehy insists.

Humankind is made "in the image of God" according to the Scriptures—an exalted standing of terrifying possibilities when corrupted, of glorious hope when honoured. Theology seeks to view reality "from the viewpoint of eternity," while centred around what it confesses as the turning-point of time, the Lord Jesus Christ. Like physics, and all the sciences, theology thereby involves people. When doctrine is expressed in terms of precise formulae that dehumanise, theology denies its calling.

In the final chapter, Sheehy lists "three key ingredients we need in order to face the challenges of the future: the ability to ask good questions; a culture of curiosity; and the freedom to persist" (p. 271). Not a bad message for theologians, one that reminded me of Jesus' approach to the disciples.

Sheehy describes herself as a "communicator of science," and by the evidence of this book, she is an excellent example. It will no doubt be read by many who are interested in science. I have no hesitation in encouraging ISCAST and non-ISCAST folk to join them in doing so.

Charles Sherlock
July 2022

Graeme Finlay: *Evolution and Eschatology: Genetic Science and the Goodness of God*

Eugene, Oregon: Cascade Books, 2021; 218 pages.
ISBN-13: 9781666704570.

This is an ambitious book, ranging from Genesis 1 to evolution of the placenta, the developing brain, immunity, and on to created histories.

Running through every aspect is genetics, with thoughtful and informed theological commentary throughout. Graeme Finlay is well-equipped to write on such matters, having degrees in theology and science. For many years, he has been a cancer researcher in the Department of Molecular Medicine and Pathology at the University of Auckland. He has written extensively on science-faith issues, and is currently a Project Director at NZCIS (New Zealand Christians in Science).

At first sight, this book could look forbidding to those without a reasonable grasp of modern genetics. That is not a criticism as such, since there are some things that require detailed insight, and that need to be wrestled with. What it does show is that Finlay's scope is vast, and that he is guided by profound theological insights as well as by a deep Christian commitment. Consequently, this is not a book for those who would score cheap shots in the well-worn evolution-creation sphere, but it is for those who want to grapple with hard issues, and be led by an expert in genetics with a very firm grounding in Christian theology.

The respective contributions of science and theology come through repeatedly, as each of their spheres of activity are outlined, pointing out their essential tenets and their corresponding limitations.

The first chapter alone on Genesis and the beginning of things is a harbinger of what is to come, as Finlay lays the groundwork for clear thinking on the status of human beings, the *imago Dei*, human dignity, and their evolutionary connections. This is well informed and more than adequately referenced. In some ways, it stands on its own, and is worth studying by itself.

The chapter on the evolution of the placenta may come as a surprise, since rarely does the general public think much about this organ, especially from a Christian perspective. Nevertheless, there are many riches here as the significance of chance events and randomness emerge. Finlay argues that random gene mutations were essential features of placental development, contributing to the functionality of the placenta in its lengthy developmental period. This, in turn, has made possible the developmental characteristics of the human brain underlying God's purposes for human beings. He rightly comments that "the postulate that evolutionary change is the bearer of God's purposes, seem to sit uneasily with each other" (p. 41). This perceptive insight leads to, what for many is, a startling conclusion: that reality is permeated with randomness and unpredictability, and that God achieves his ends in history despite and through the randomness that characterises the behaviour of his creatures (p. 48). Coming back to the placenta, Finlay reminds us of the essential inter-relationship between parent and unborn child: "the advent of the placenta has provided the conditions enabling prenatal parenting" (p. 56). This is a salutary perspective that is usually overlooked by those who regard the embryo and foetus as having absolute value in isolation of any considerations of their embryonic or foetal environments.

In turning to the developing brain, Finlay again seeks to follow genetic changes during brain evolution, leading him to conclude that there is compelling evidence of human descent from ancestors with recognizable genetic characteristics. Myriad tiny incremental steps have provided, he writes, genetic specifications underlying the complexity of the human brain (p. 69). These emphasise our embeddedness in the materiality of biological history. However, he is no materialist, because he is forthright with his assertion that lasting meaning is closely tied in with humans as personal agents.

Finlay, therefore, is careful to avoid genetic determinism as he emphasizes the central significance of community, our interaction with other personal beings, and the quality of our nurture as social beings. Much of this discussion veers away from genetic input, since it is less dependent upon an understanding of genetics. This is no bad thing, and it does underline an important point: that genetics is not everything. Nevertheless, Finlay is aware of some of the central drivers within neuroscience, all of which point to the interplay of biology, socialisation, and neuroplasticity. He concludes: "The structure of my brain ... has been formed by God's knowledge of me and by my painfully clouded and incomplete knowledge of him." (p. 90).

The chapter on immunity brings Finlay back to home territory, with his description of innate and adaptive immunity. Here he touches on the manner in which natural selection is distasteful to some because of its dependence upon free randomness, and hence its apparent incompatibility with a purposive God (p. 103). Against this, he contends that in each of us as individuals, natural selection serves as a powerful mechanism for generating new immunological capacities. This is because immunity develops from the ongoing interaction between genes and their "indefinably complex environment" (p. 106). Hence, immunologically we are not self-sufficient, autonomous gene machines.

It is fascinating the way in which he sees the complex and manifold interdependencies of the immune system as paralleling the interdependencies of the body and of life in the Christian church. It could be argued that he pushes this parallel between the immune system and the Christian community too far, but it cannot be denied that, no matter how speculative it is, it serves to open up new areas for contemplation and creative thought.

The significance of the book's title emerges in the final chapter on "created histories." In this he argues against the notion that God micro-manipulates the world and human beings through mutations, especially through those that cause cancers. Rather, he puts forward an understanding of evolutionary process as history that recognizes its ambiguity, contingency and genuine ontological freedom (p. 116). More specifically, he contends that random, autonomous process (chance) leads evolution along particular trajectories, so that, in his view, phylogenetic history invites a teleological interpretation. In arguing like this, his strong Christian commitment shines through, even as his interpretation will be one among many.

I am grateful for the manner in which Finlay has brought his deep genetic expertise, especially in the cancer arena, to bear on evolutionary matters. He has been prepared to look deeply into evolutionary territory and has not been afraid to face up to the challenges this has for theology. The destructiveness of cancers poses enormous problems for Christians with their picture of a loving God who seeks only their best. Finlay writes: "the achievement of something resoundingly good may be attained only at the cost of undesirable side effects" (p. 133). As he grapples with this dichotomy of good and evil in the world, he finds himself walking through territory already surveyed by countless other Christian thinkers. He is only too aware that the genetic mechanisms of biological evolution and tumour evolution are the same. How then can we affirm the creative purposes of God in the former, but deny them in the latter? He is prepared to confront this dilemma, not shirking the immensity of the issue.

I admire Finlay's willingness to do this. In doing so, he delves into the significance of the "fall," human sinfulness, the incompleteness of biological and human stories, the openness of history, and the completion of all stories in Jesus the Messiah. Inevitably, there is much in his perspectives that is open to debate. However, what is important is that his views are put forward with humility and integrity. He concludes with the phrase: "Nothing in evolution makes sense except in the light of eschatology" (p. 158). This will not be welcomed by everyone, but Finlay has put forward enormous ideas that require ample further discussion by those seriously committed to dialogue at the science and faith boundary. This book is not light reading and its ready acceptance of evolutionary concepts will not be welcomed by all theologically conservative Christians. Nevertheless, there are riches in store for those prepared to engage with someone who takes theology and genetic science seriously.

D. Gareth Jones
University of Otago
September 2022

www.ingramcontent.com/pod-product-compliance
Lightning Source LLC
Chambersburg PA
CBHW071958290426
44109CB00018B/2059